Mineral Resources and Energy

Series Editor
Alain Dollet

Mineral Resources and Energy

Future Stakes in Energy Transition

Olivier Vidal

First published 2018 in Great Britain and the United States by ISTE Press Ltd and Elsevier Ltd

ISTE Press Ltd
27-37 St George's Road
London SW19 4EU
UK

www.iste.co.uk

Elsevier Ltd
The Boulevard, Langford Lane
Kidlington, Oxford, OX5 1GB
UK

www.elsevier.com

Notices

Knowledge and best practice in this field are constantly changing. As new research and experience broaden our understanding, changes in research methods, professional practices, or medical treatment may become necessary.

Practitioners and researchers must always rely on their own experience and knowledge in evaluating and using any information, methods, compounds, or experiments described herein. In using such information or methods they should be mindful of their own safety and the safety of others, including parties for whom they have a professional responsibility.

To the fullest extent of the law, neither the Publisher nor the authors, contributors, or editors, assume any liability for any injury and/or damage to persons or property as a matter of products liability, negligence or otherwise, or from any use or operation of any methods, products, instructions, or ideas contained in the material herein.

For information on all our publications visit our website at http://store.elsevier.com/

British Library Cataloguing-in-Publication Data
A CIP record for this book is available from the British Library
Library of Congress Cataloging in Publication Data
A catalog record for this book is available from the Library of Congress
ISBN 978-1-78548-267-0

Printed and bound in the UK and US

Contents

Foreword

This book and the geophysical works of Olivier Vidal are part of the vast field of fundamental re-writings on economic science. Economic analysis continuously reflects upon the fact that capital and labor alone are sufficient to generate prosperity. However, two elements are missing from this truncated perception of the world: energy and material.

Capital without energy is like a sculpture and portfolios of financial assets. Human labor without energy is like the inactivity of a corpse! Deprived of material, either one is no more than an abstract idea. In other words, the economy, for more than two centuries now, has relied upon a virtual world, populated by morbid ideas: such as corpse-like structures and virtual speculation.

Olivier Vidal's contribution reintroduces in our understanding the contemporary crisis that strikes the economies of the planet, the two physical ingredients without which no life would be possible: material and energy. In doing so, the following pages highlight some crucial truths, these being the metamorphosis that we must introduce into society in the shortest possible time, by moving from an inherited economy based on industrial revolutions – essentially hydrocarbon resources and the unrestrained plundering of mineral resources – towards a sensible economy of renewable resources and materials. The first of these truths is that the world will lack certain materials in the decades to come. Lacking does not mean that one morning we will find ourselves without any available mineral resources. As Vidal points out, history shows that mineral reserves have steadily increased in recent decades by way of technical progress and their higher market value, making it more attractive to explore mineral stock limits in the Earth's crust – a stage that the 1972 Meadows Report did not anticipate.

Lacking means that our mineral consumption has increased exponentially since the 19th Century; there will inevitably come a time when the quantity of material available for extraction will no longer be able to increase owing to the speed of extraction. Olivier Vidal's work indicates that the peak of copper extraction could occur before 2060. A long time ago in Ancient Cyprus, it was simply a matter of bending down and finding copper beneath your fingertips. The year 2060, based on a geological time scale, is today. For the economic cycle this sounds like tomorrow. For the financial markets, it is a worthless eschatological horizon and without a doubt this is where the major problem of financialized societies lies: they have great difficulty in understanding what economists like to call the "long term", but which only exists in light of the dematerialized time in which their dreams are awakened. The consequence? A significant part of the assets upon which the wealth of economies is supposed to be based actually becomes a financial bubble. In the same way that bankers and insurers are gradually becoming aware of the fact that carbon risk threatens to plummet the value of several trillion assets, whose price does not take into account the pressing threat of climate change. Likewise, many industrial assets are valued today as if the cost of energy and material and their operation and maintenance were negligible. What will be the point of having gadgets if there won't be enough spare parts to repair them, let alone natural resources available to replace them?

The second truth that will be revealed to the reader is that the scarcity of natural resources is not uniform. For example, there is still a lot of lithium on the planet, so batteries have a good future. As a result, economic analysis must now take into account Mendeleev's entire periodic table if it wants to understand what the future will hold. Which metals are we able to dispense with today? Which metals with industrial uses do we have reasonable hope of finding substitutes for in the coming decades? This is where the third truth of the book comes into play: the energy–material nexus is the focus of a large number of physical constraints that affect human activities. Due to the fact that it requires more and more energy to extract mineral resources from the subsoil, the density of the reserves we use decreases. However, you still need metals to harness energy! Thus, the infrastructures associated with renewable energies are more eager when it comes to copper extraction than with the "dirty" oils which are at our disposal. If we are not careful, we could find ourselves in a deadly bottleneck: one that would inevitably close in on us if we squander too much copper. We need to be able to extract enough of it in order to install and maintain the infrastructure required to replace coal-fired power plants and oil rigs. Olivier Vidal's central message is not to obtain complete restraint, which – to those who prefer the sophisticated dream of dematerialized corpses-traders compared to reality – raises the fear of a return to the "Stone Age". Vidal's message is a call for a reasoned usage of natural resources in order to respect our anticipated long-term needs.

Is recycling not the best solution to this foreseeable impasse? It is obviously part of the solution. Nevertheless, Europe has been slow in recycling its waste. In addition, its industrial deployment requires us to think *ex ante* about consumer objects which are easy to recycle: the shift towards nanotechnology leads to the recycling of technological devices becoming more and more complicated and expensive...in terms of energy. Again, a bottleneck is to be feared if we do not intelligently anticipate the *material* reality of recycling. Finally, recycling is vital but it does not liberate us from exponential law. Supposing that like the city of San Francisco we managed to recycle 80% of our waste in one year, after 10 years we will have only retained 10% of this precious material. In other words, the *decline* in the rate of recycling towards zero is unfortunately as fast as the growth of our consumption towards an upper limit that only a handful of economists believe can be postponed *ad libitum*.

The economy which is outlined over the following pages has nothing to do with corpses playing the markets. Instead, it consists of seeing human society as a vast living metabolism. Like any living organism, it takes natural resources, converts them into "effort" (whose GDP is only a very approximate monetary measure) and rejects all which it has taken and not used as effort in the form of waste. An economy is nothing more than what Ilya Prigogine calls a dissipative structure, a converter of material and energy, maintained at a distance from the thermodynamic equilibrium by way of the flow of natural resources that it borrows and converts. So that this said metabolism can maintain its physical fabric (but also its degree of structural complexity, of which entropy is a measurement) it *must* maintain, at least, the flow of what it harvests. Otherwise, it will converge towards thermodynamic equilibrium: examples being when a hot drink cools to the temperature of its environment and a building when it is no longer maintained and after several thousand years it will no longer be distinguishable from the soil upon which it was built. To tell the truth, if the conventional economy does not listen to Olivier Vidal's lesson, this is the future it promises to our societies: a waking nightmare that will soon become reality and a vision that our societies will soon resemble crumbling museums.

What do we need to think about in terms of a living economy? Do we need to think about metabolisms that produce daily entropy in order to maintain and develop an internal structure? Firstly, there are data. We are sorely lacking in quantified resources that constitute material, the sampling flows we make every day on limited resources, the international mineral trade and the dependency of nations on mineral extraction. Olivier Vidal's contribution is the first step in the right direction. Secondly, we need a renewed intelligence of the thermodynamics of material and energy flows that help us survive. Once again the last chapters of this book provide valuable lessons: the prey–predator dynamic, which has already proved fruitful in biology, is an essential element in this new perspective offered by Vidal. We can wager that this perspective will very quickly become one of the cornerstones of tomorrow's living economy. It is at this price that the latter will finally become "scientific".

Gaël GIRAUD

Research Director at CNRS

Chief Economist at Agence Française de Développement

Director of Energy and Prosperity Chair

1

Framework and Challenges

1.1. The strong growth of mineral resources and energy consumption

The industrialization of developed countries has led to an increase in productivity of all human activities and to the replacement of labor by humans and animals with that of machines fueled by fossil fuels. This industrialization, which began in Europe in the 19th Century and which is currently taking place in developing countries, brings about many social, economic, demographic and technological changes. These developments at the global level are illustrated by the apparent exponential growth of all indicators of prosperity, human activity and its impacts [STE 14]. The growth of the population and its urban proportion, the increase in economic activity, average income and the standard of living lead to an equally exponential growth of energy and raw material needs (Figure 1.1).

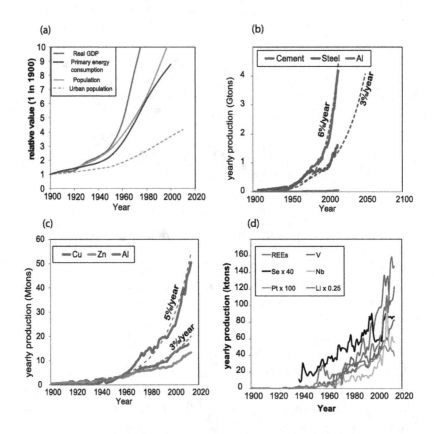

Figure 1.1. *Historical evolution of the various indicators of prosperity and human activity: (a) standardized trends in world population, urban population, GDP and primary energy consumed; (b) to (d) changes in annual production of cement and various metals. For a color version of this figure, see www.iste.co.uk/vidal/energy.zip*

The human race is currently using mineral resources at an unprecedented level with about 70 billion tons extracted annually [WIE 15]. Aggregates, cement, sand, metals and industrial minerals are at the forefront as they are used to build urban infrastructure, consumer and production goods, including energy production and use. The industrial and economic development of a country can be broken down into two main stages:

– The first stage includes the construction of basic infrastructure such as housing, infrastructure, transport and communication, production, transmission and distribution of energy, heavy industry, etc.

Urban density is also a contributing factor to residential consumption [SAF 07]. The transition from a rural to an urban lifestyle leads to the transition from biomass energy to fossil energy. Energy efficiency has generally increased, but its uses are more demanding and diversified, such as the use of air conditioning for instance. Between 1990 and 2012, the proportion of the population of China residing in cities rose from 26 to 52%, and according to the World Bank, residential energy consumption increased fourfold over the same period. According to the International Energy Agency [IEA 16], energy consumption for the residential sector increased by 7.7% per year between 1998 and 2012 and per capita energy use increased at the same rate between 2001 and 2012, while that of OECD countries fell by 0.8% per year. This development phase consumes mainly "structural" raw materials such as concrete, steel and other base metals such as copper and aluminum. After a strong period of growth, annual consumption stabilizes or begins to decline [BLE 16], with an increase in the standard of living from about US$ 15,000-20,000 of gross domestic product (GDP) per year per inhabitant for steel, US $25,000 for concrete [DAV 14] and US$ 30,000 for copper[1]. According to the International Monetary Fund, a significant number of countries with large populations currently have a real GDP per capita of less than $ 15,000, including China, Indonesia, the Philippines, India, Pakistan and some African countries. As in the case of developed countries, their emergence is inevitably associated with an increase in the consumption of raw materials used

---

1 https://www.fxstreet.com/analysis/economic-monthly-report/2015/06/22.

for the construction of buildings and basic infrastructure. This is illustrated by the strong growth since the 2000s, as observed in Figure 1.1, of steel and concrete (+6%/yr), aluminum (+5%/yr), copper (+3%/yr), boosted by the very rapid emergence of the BRICs (Brazil, Russia, India, South Africa and of course China). Chromium (+5%/yr), manganese (+6%/yr), nickel (+5%/yr) or zinc (+4%/an) are following the same trends.

The increase in world population to nine billion by 2050 and the rise in the standard of living of the poorest countries, which accounted for 85% of the world's population in 2005 and consumed only 10% of the most common metals (Figure 1.2), will undoubtedly increase the need for mineral resources and raw materials until the middle of the 21st Century. It is estimated that the maximum stock of iron and steel in developed societies is about 10 t per inhabitant [RAU 09, MUL 11, WIE 15]. Beyond this, the level of consumption corresponds to that needed to replace and maintain the stock level, i.e. about 500 kg/inhabitant/yr. By comparison, the average world iron and steel stocks per capita is estimated to be about 2.7 t and the annual global average consumption is about 200 kg/inhabitant/yr [MÜL 06, ALL 12, MÜL 11]. To move from a current stock of 2.7 t/inhabitant for a population of 7 billion to a stock of 10 t/inhabitant for 9 billion individuals, 71 Gt of iron and steel would have to be produced, assuming there is no loss. This would represent 85% of known reserves[2]. For this evolution to take place in 35 years, the average steel production must reach 3 Gt/year in 2050, twice the current production rate. According to Grädel [GRÄ 11b] and Grädel and Cao [GRÄ 10], the total quantity of metals to be produced by 2050 and the flux of used metals could be 5 to 10 times the current values.

---

2 USGS Mineral Commodity Summaries, January 2017, https://minerals. usgs.gov/minerals/pubs/commodity/iron_ore/mcs-2017-feore.pdf.

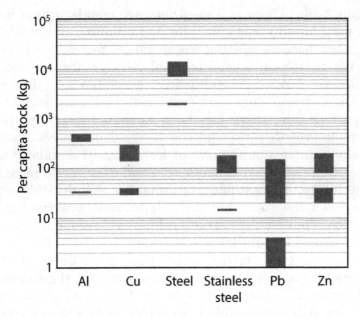

Figure 1.2. *Estimate of the stock per capita of different metals. The blue bars represent the more-developed countries (860 M people); the red bars represent the less-developed countries (5,620 M people). Data source: [UNE 10b]. For a color version of this figure, see www.iste.co.uk/vidal/energy.zip*

– In a second stage, when "structural" raw materials needs stabilize, developed countries are moving towards advanced technology, particularly in the sectors of electronics, automation, information and communication, energy, production and supply chains. Advanced technologies require new raw materials and mineral resources. At the beginning of the 20th Century, metal consumption was limited mainly to iron, copper, lead and zinc and silver, which had the desired basic physical and chemical properties such as hardness, malleability, corrosion resistance, density, conductivity or electrical resistivity. Advanced technologies use many additional properties, including electronic structure, catalytic, quantum or semi-conductive properties specific to almost all elements of the periodic table. There is a need for rare metals, which have been used for only a few decades and produced in quantities lower than base metals, but whose annual production growth can reach record levels (around 10%/yr for antimony, beryllium, cobalt, gallium, germanium, lithium, molybdenum and some rare earth metals, Figure 1.1). Over the past decade, there has been increased focus on these enormous growth rates and the heavy dependence of most wealthy countries on imports of technological metals. While it is certain that their consumption will continue to increase in the future, a quantified assessment is still difficult, as it depends on the rapid but difficult to predict technological evolution. These, unlike structural mineral resources, cannot be substituted for all their applications. Another difference is that technological metals are often by-products of the extraction of a major metal that accounts for the financial profitability of mining. If the demand for a by-product metal increases, it is not possible to increase production because it is the demand for the supporting metal that determines the level of production.

The increase in consumption of raw materials and energy is the result of three main factors: population growth, industrial development and an increase in the average standard of living. These developments are accompanied by a massive increase in the exploitation of increasingly diversified mineral raw materials, first with steel, cement and copper, light metals (Al, Mg, etc.), then rare metals and very pure materials for advanced technology, such as rare-earth elements (REEs) or silicon for example. Between 1940 and 2010, the world's population increased threefold, while during the same period consumption increased fiftyfold for cement, eightfold for steel, manganese, copper and zinc, twenty-fivefold for platinum and much more for all the elements used in advanced technologies. This growth is not expected to decline globally in the decades to come, as we are in a very dynamic phase that is the result of the simultaneous growth of the Chinese economy and the rapid emergence of advanced technology. The rest of Asia, India and Africa have the opportunity to follow the Chinese evolution and current projections indicate that the cumulative amount of metals to be produced over the next 40 years will exceed the cumulative amount that has been produced to date.

1.2. Mineral resources and energy in the context of energy transition

The invention of the steam engine was at the origin of the Industrial Revolution, worldwide trade of materials and the increasing demand for mineral resources, including metals. The use of copper made it possible to increase the efficiency of boilers. Alloyed with tin, it allowed the production of bronze, used for the bearings needed for many mechanical devices, including machine tools. Steel producers then noticed the profits associated with the addition of chromium and manganese. The invention of the combustion engine was a second revolution that generated the production of a whole infrastructure for the production and transport of hydrocarbons, its processing by the petrochemical and gas industry, but also the different means of locomotion and personal vehicles produced over a century. This rapid evolution has been possible due to the access to cheap and abundant fossil energy. This situation is likely to change because the emissions of carbon dioxide, particulate matter and other combustion-related components have worrying environmental consequences, whether on climate and global warming or air quality in large urban areas. The Paris Agreement of COP 21 (the 21st meeting of the Conference of the Parties), which plans to achieve "carbon neutrality" sometime between 2050 and 2100, will involve a massive reduction of carbon dioxide emissions and an in-depth revision of the existing global fossil energy-based system. This will require a thorough overhaul of the existing energy system, which is mainly based on fossil fuels and coal worldwide, and possibly a significant reduction in the amount of energy used. Transition to less carbon-emitting energy requires new materials. This is true both for the nuclear sector and for the generation, storage and distribution of electricity generated from renewable sources. In any case, new infrastructures have to be built using "structural" raw materials, such as steel, aluminum or copper, but also rarer metals: neodymium, praseodymium and dysprosium in super-magnets of certain wind turbines, tellurium, indium, gallium selenium for photovoltaic thin films, lithium and cobalt for batteries of hybrid or electric vehicles, etc. All of these raw materials require a lot of energy to be produced: at present, about a quarter of the world's energy consumed by industry is used for the production of steel, cement and aluminum alone. It can therefore be envisaged that the energy transition will at least initially lead to the overconsumption of fossil energy and will also be accompanied by an overconsumption of metals.

Ensuring the transition to low-carbon energies and energy sobriety is certainly an important issue for the 21st Century, but it is not the only one. This evolution will have to occur during a period of increasing urbanization, which is also associated with increasing needs for raw materials and energy: cities consume about 80% of the world's energy production and produce 75% of greenhouse gas emissions. If we imagine a 50% reduction in global energy used by the raw materials sector ([ALL 10], Intergovernmental Panel on Climate Change, 2011; United Nations Foundation, 2007[3]) and a doubling of the demand for raw materials for developing countries, a 75% reduction in energy consumption would be required for the production of raw materials currently dominated by steel, cement, aluminum, paper and

3 United Nations Foundation (2007), Realizing the potential of energy efficiency: targets, policies, and measures for G8 countries. Washington, DC, http://aceee.org/files/pdf/conferences/mt/2008/plenary_chandler_supp.pdf.

plastics. Some studies suggest that this will be difficult to achieve [GUT 12] and in any case, a sharp reduction in greenhouse gas emissions in the context of increasing demands for raw materials leads to new constraints which must be taken into account and anticipated. This concerns the conditions and technologies of energy transition, which are closely linked to those of the production of raw materials. In 2010, 1.4 billion people, or 20% of the world's population, did not have access to electricity [SAT 11] and 2.7 billion people, or 40% of the world's population, cooked using biomass in an unsustainable manner [IEA 10a]. Ensuring secure access to electricity and maintaining energy vectors for everyone and for our daily uses is not just about the availability of the primary source, but also about the entire infrastructure necessary for its capture, processing, storage, distribution and use. The issues related to energy and mineral raw materials are therefore closely linked: the production of raw materials requires energy (which is discussed further in Chapter 3) and the production of energy requires raw materials (Chapters 4, 5 and section 6.1).

1.3. Towards a shortage of mineral resources?

If long-term growth trends in population and prosperity continue as they did in the past, the overall supply of metals to be produced by 2050 could be 5 to 10 times the current level [GRÄ 11]. The Earth is an immense reservoir of elements whose accessibility is limited by our ability to identify and access the resource when it is deep and hidden, to anticipate geopolitical risks and to control the environmental and social impacts associated with extraction. The exponential increase in demand for most fossil resources over the past century has raised the question of their "finiteness". Possible restrictions on supply would seriously affect our societies, as humanity has already experienced [TAI 88, TUR 09, CHA 97, MOT 14]. Different studies suggest that the production of many fossil resources has peaked or will peak in a foreseeable future (see [LAH 10, KER 14, NOR 14, SVE 14a, SVE 14b] for the case of copper). This research argues that current trends in population growth and consumption of fossil resources are unsustainable. In the early 1970s, Meadows *et al.* [MEA 72] expressed the same fears and anticipated the exhaustion of many resources within a short period of time. This has not been the case, as improved technologies have made it possible to discover new deposits and to exploit those that could not be used at reasonable cost with past technologies. The known reserves of many metals are more significant today than those known 50 years ago, and the time before depletion of known reserves has remained relatively constant over the last century. Nevertheless, it is certain that the exponential growth in consumption of raw materials will not continue indefinitely because our planet is finite in size and because infinite growth is not necessary. Demand is expected to stabilize when all countries will have reached a homogeneous level of development and a GDP/inhabitant corresponding to the level of saturation of raw materials in society. At this stage, recycling could be the main source of raw materials. Before reaching this stage, the increase in recycling and in general in the circular economy will not be sufficient to meet the needs. Indeed, we can recycle only some of the consumer goods and equipment that were created several decades ago, when the consumption of raw materials was much lower than current needs. During the growth period, increasing the production of primary raw materials poses several questions that are not new but are still relevant. They concern the possible depletion of resources before reaching the saturation stage and our ability to maintain a doubling of the production every 15 to 20 years, which was possible in the past. This topic is discussed in section 6.2 using dynamic modeling.

2

General Information on Mineral Raw Materials and Metals

2.1. Relocation of primary production and dependence of industrialized countries on imports

Although the last century was marked by a sharp increase in the consumption of fossil energy and metals, the deflated price of base metals on average has experienced much more moderate fluctuation (Figure 2.1).

The price, in US$ constant currency, of 17 commonly used mineral raw materials remained relatively constant throughout the 20th Century, before rising sharply in the early 2000s, under the effect of the spectacular growth of the Chinese economy. The stability of prices in constant currency, despite the massive increase in consumption, is linked to the technological advances that took place since the beginning of the 20th Century, which increased the productivity and made possible the exploitation of less concentrated reserves. The stability of prices is also linked to a relocation of production to lower-GDP countries, where production costs are lower.

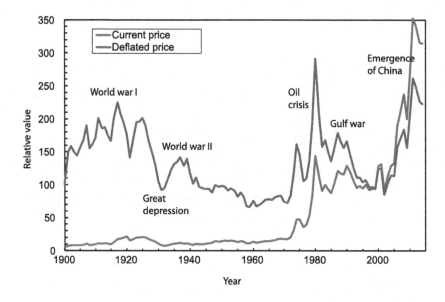

Figure 2.1. *Evolution of the average annual price of 17 raw materials (Al, Be, Fe, Cr, Co, Au, Pb, Mn, Hg, Mg, Mo, Ni, Pt, Ag, Sn, Tn, Zn), in US$ constant currency (base 100 in 1900) and current currency (base 5 in 1900). Data source: U.S. Geological Survey: http://minerals.usgs.gov. For a color version of this figure, see www.iste.co.uk/vidal/energy.zip*

The collapse of the USSR and the resulting decrease in geopolitical tensions, coupled with moderate growth of consumption between 1970 and 2000, resulted in a lack of political interest in mineral-related issues in most western countries. Until the early 2000s, the logic of deregulation of markets was supposed to favor the conditions necessary for supply in response to a growing demand for raw materials at the global level. Often due to the lack of regulation on the environmental and human impacts linked to the exploitation of resources, some countries have obtained a quasi-monopoly position in production. This is the case for China for rare earth metals and other critical elements such as antimony, gallium, fluorite, germanium, graphite, indium or tungsten. This is also the case for South Africa (platinoids), the Democratic Republic of Congo (cobalt, tantalum), Brazil for niobium, and the United States for beryllium.

The case of the rare earth metals is emblematic. Contrary to what their name suggests, not all rare earths are rare, with world reserves of rare earth oxides estimated at 115 million tons in 2011 [1] for an annual consumption of 150 to 200 thousand tons, by 2015. Prior to 1965, rare earth metals were mined in South Africa, Brazil, India (10 kt/yr), and then from 1965 to 1985 in the USA (50 kt/yr). From 1985, China took the production monopoly to more than 95%, with a production of >100 kt/yr, while having an estimated 35% of the world's reserves. The main reason for this is economic: the rare earth metal deposit of Bayan Obo, in Western Mongolia, was initially mined for iron and the rare earth metals were a bonus, with free mining.

Other reasons are environmental and social: the exploitation techniques that were used for a long time did not take into account the social, health and environmental costs of extraction. Extremely polluting technologies that would not be feasible in developed countries have been adopted, resulting in pollution by actinides (thorium, uranium and radium) naturally associated with rare earth metals. In Western countries, mining waste from the extraction of rare earth metals is similar to radioactive waste and is subject to binding regulations the compliance to which increases the extraction costs. This is one of the reasons that led these countries to limit or stop production.

---

1 Mineral Commodity Summaries 2014, United States Geological Survey, Reston, USA, available at: http://minerals.usgs.gov/minerals/pubs/mcs/2014/mcs2014.pdf.

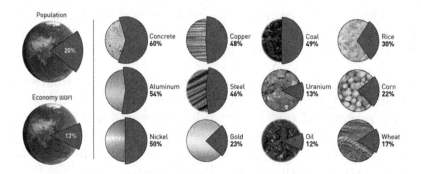

Figure 2.2. *Proportions of raw materials consumed by China.
Source: http://www.visualcapitalist.com/china-consumes-mind-
boggling-amounts-of-raw-materials-chart/. For a color version
of this figure, see www.iste.co.uk/vidal/energy.zip*

From 2005 China announced that the environmental consequences of its production had to be addressed and in 2010 the Chinese government announced that the environmental problems were such that they had to reduce their production and exports. It is easy to imagine that other strategic and political reasons have motivated this decision, but whatever the causes, the decline in exports has been a wake-up call and in a few years, hundreds of rare earth mining projects have emerged outside China. This confirms that production monopolies are not a geological fatality. There is an important potential for discovery of metal deposits in countries where investment in modern mineral exploration has been weak or non-existent in the past. The European Union, whose mineral heritage is only partially known, also has potential [RAB 12]. However, the challenges of primary extraction in rich countries are not only related to the presence or absence of resources. Extraction also requires the combination of acceptable social, economic and regulatory conditions. This is fueled by the vast improvement in extraction and production technologies since the 1980s, which can make possible and acceptable what was not possible 40 years ago from an economic and environmental point of view.

While the shortage of raw materials never materialized in the 20 Century, there have been numerous strains on several occasions. The economic paradigm that has prevailed so far to ensure the supply to rich non-producing countries has been the existence of mining activities in poorer countries and the availability on a supposedly undistorted market of raw materials that can be imported for conversion into high value-added products by the domestic industry. This model worked well in the 20th Century but is now obsolete, notably because of China's very rapid development as a major player in the global mineral industry, with a policy of acquiring rights (Figure 2.2) in order to ensure the production of high value-added industrial goods in its territory.

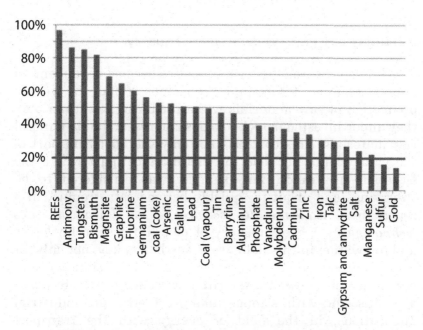

Figure 2.3. *China's global share of mining or metallurgical production in 2011. The red bar represents the proportion of the Chinese population in the world population. Data source: [REI 13]. For a color version of this figure, see www.iste.co.uk/vidal/energy.zip*

Figure 2.3 shows the proportion of China's mining and metallurgical production for the 28 mineral commodities in 2011 when it became the largest producer in the world. Production is not intended to meet the needs of the downstream industries of rich countries but to supply downstream industry established in China. The desire to no longer export raw materials with low added value, but to develop high value-added industrial endorsements is shared by a growing number of producer countries. This is reflected in the multiplication of tariff and non-tariff barriers restricting the free international trade of mineral raw materials. In OECD, in 2014, more than 9,000 individual measures[2] were taken by 69 countries on export restrictions on mineral raw materials.

2.2. Uncertainty and vulnerability of supply

While non-commodity rich countries became dependent on imports (Europe, for example, consumes 20% of the world's production of raw materials while producing less than 5%), they under-invested in technologies and research into mineral raw materials. This concerns not only the upstream part of the value chain, but also the metallurgical part where the situation in France, for example, was considered to be "worrying" and about to "quickly become catastrophic" in a report by *l'Académie des sciences et de l'Académie des technologies*[3]. A good mastery of metallurgical technologies and innovation in this field are nevertheless key, not only for primary production but also to optimize the recycling capacity of metals and in general the circular economy. Metallurgy is also a sector with strong links with strategic industrial applications, in the field of energy with the transport

---

2 Restrictions on Exports of Raw Materials online database, OECD, Paris, France, 2012, http://qdd.oecd.org/subject.aspx?subject=8F4CFFA0-3A25-43F2-A778-E8FEE81D89E2.
3 http://www.france-metallurgie.com/pour-la-metallurgie-en-france-l'academie-des-scien es-et-l'academie-des-technologies-s'engagent.

industry, nuclear and thermal energy transformation industry, as well as eco-design[4].

The problems specific to the mineral industry are numerous and varied. They include the gradual decline in the grade of the deposits, the increase in investment required for production and production costs, difficulties in accessing resources, or the lack of discovery of high-quality deposits. Large deposits of highly concentrated minerals in major OECD producing countries with the conditions for responsible mining have likely been discovered. This does not mean shortage, but the gradual depletion of enriched and near-surface areas requires the exploitation of less and less concentrated deposits. Some waste from previous mines is as concentrated in metals as some large deposits currently in operation. These developments lead to numerous hazards that can affect the industrial sectors and impact the importing countries and their industries that use raw materials:

– political and institutional hazards: absence or change of state policy, lack of coherence of regulations (environmental, fiscal, mining, labor etc.) with an impact on the mineral industry, weaknesses of the institutional framework, bad governance, lack of transparency, corruption, introduction of tariff and non-tariff barriers restricting the export of the production of such by-products, infrastructure with poor water and/or electricity supply and/or other inputs into the transport infrastructure network;

– economic effects: volatility in commodity prices, decline in grades, increase in the cost of energy and equipment, decrease in demand (technological changes and regulatory developments), tariff and non-tariff barriers affecting the materials market's use of monopoly positions as a geopolitical

4 La Métallurgie en France, une nécessité d'innovation, Report by the French Ministry of Economy, Finance and Industry, 2005, https://archives.entreprises.gouv.fr/2012/ www.industrie.gouv.fr/enjeux/metal.pdf.

instrument, cost of investments required for the industrial project, cost of guarantees required by the State, insurance;

– environmental effects: depletion and/or pollution of water resources, soil pollution, erosion; air pollution, toxic discharges, risks linked to the stability of land affected by mining, water inflows, risks of breaking up mining waste or metallurgical slag, exceptional climatic events (drought, floods);

– geological effects: earthquakes, tsunamis, volcanic activity, landslides, subsidence;

– technical contingencies: disappearance of the technical skills required for the mineral industry, poorly estimated reserves, technological limits for the concentration of diluted substances (primary materials or end-of-life products);

– societal effects: unmanaged cultural differences between the mining operator and populations impacted by mining development, opposition to the development of the mineral industry, lack of reliable information and data relating to the mineral industry, strikes, civil wars.

These effects are just as important to consider as the availability of resources. Ali *et al.* [ALI 17] show that one third of the current world primary copper reserves are located in countries whose governance is deemed incompatible with the International Resource Governance Index published in 2013 by the Revenue Watch Institute[5]. This index, which measures the quality of governance in the oil, gas and mining sectors of 58 countries producing 85% of oil, 90% of diamonds and 80% of copper in the world, assesses the quality of four key components of governance: institutional and legal framework, reporting practices, guarantees and quality control, and environment. Eleven producing countries, or less than 20% of the countries studied, have satisfactory standards of transparency and accountability. This creates an additional

5 http://www.resourcegovernance.org/sites/default/files/rgi_2013_Eng.pdf.

supply risk, especially if raw materials are produced in fewer countries. Awareness of the risks and the resulting vulnerability has motivated many studies, analyses and strategies. Economic, social, environmental and industrial issues are of prime importance and it should be considered that in the future, the supply conditions of rich countries may no longer be ensured by production outside their borders as this has been the case in the past.

2.3. Recycling of waste

Waste recycling is an environmental imperative and could become the major source of metals in the second half of the 21st Century (see Chapter 6). In the case of metals, there are two main types of waste: end-of-life products and mining and smelting waste. In both cases, potential deposits are enormous even though regulatory changes are increasingly restrictive and end-of-life products and the metal alloys they contain are becoming increasingly complicated and difficult to recycle.

2.3.1. *Primary production waste*

The type, quantity and properties of mining waste produced in different mines will vary depending on the resource exploited, production technology and geology of the site being mined. Technological advances and regulatory changes have resulted in positive waste management developments over the past 10 to 20 years [LOT 11, LOT 12] and an appropriate waste management plan is now required before issuing an exploitation permit. Different guidelines on waste management, including post-mine closure, have also been developed at the international level. The term "waste" refers to material derived from primary production, which has no economic value at the time of production. This waste includes[6]:

– rock waste, which is aggregate with a high content of metals. This waste can be reprocessed to extract metals, or used as backfill, as building material and aggregate. There is a lot of research into the potential of extracting metals from mining waste using conventional methods as well as unconventional ones. An interesting example is BacTech Environmental Corporation[7], which was created in December 2010, using patented bio-leaching technology for the treatment of toxic arsenic-bearing tailings. The process stabilizes arsenic and oxidizes sulfides from tailings and therefore eliminates the major source of acid production. Finally, the technology recovers precious and base metals from the residues. The technology makes it possible to treat mining waste accumulated through years of exploitation by optimizing the living conditions of existing and benign bacteria for humans and the environment. Once the treatment is complete, the inert residues can be used as backfill. Another example is described by Malvoisin *et al.* [MAL 13] proposing the use of geomimetic approaches to produce iron oxides which can be used for pigments, for example, and high purity hydrogen from slag by hydrothermal reaction;

– tailings, which are finely ground rock and mineral waste products from mineral processing operations. Tailings may contain chemicals and are usually stored in the form of water-based slurry in sedimentation lagoons enclosed by dams. In the 1970s, there were 44 tailings dam malfunctions, 27 malfunctions in the 1980s and 7 in the 1990s [RAJ 05]. Manganese residues have been used in agroforestry, coatings, resin or glass and glazes. The clay-rich residues are used to make bricks, tiles and cement;

---

6 Fraser Institute 2012: http://www.miningfacts.org/Environment/How-are-waste-materials-managed-at-mine-sites.

7 http://www.bactechgreen.com/s/Home.asp.

– slag, which is a by-product from metal smelting. Slag is largely harmless to the environment and generally used as aggregates in concrete or roads. The red bauxite mud is a solid alkaline residue produced in aluminum refineries. It can be used in the treatment of wastewater or as a raw material for glass, ceramics and bricks;

– mine water from the mining sites must be treated before being released into the environment. In extreme cases, where residual sludge is rich in toxic elements such as cadmium, mercury or arsenic, it can be classified as hazardous waste and requires specific treatment [YOU 02];

– gaseous waste: gaseous waste includes dust and sulfur oxides produced during pyrometallurgy. Current technologies eliminate up to 99.7% of dust and fumes and 80-95% of sulfur oxide emissions. In Canada, for example, emissions of sulfur dioxide from metal smelters decreased by 37% between 2003 and 2010[8]. The residues may be used for the manufacture of sulfuric acid.

2.3.2. *End-of-life products*

In a world that exponentially consumes manufactured goods, the possibility of recycling metals is becoming a major issue. Recycling issues are dealt with in detail in numerous publications, including one published by UNEP [UNE 13b], which provides a comprehensive overview of the challenges. This chapter merely analyzes the issue and interested readers are invited to read the UNEP report, which is extremely well documented and refers to the main studies on the issue.

There are vast "hidden reserves"[9] of raw materials in industrialized countries. It is estimated, for example, that the city of Sydney in Australia contains 600 kg of copper per

---

8 Canada, Environment Canada, National Pollutant Release Inventory, 2012: http://www.ec.gc.ca/inrp-npri/default.asp?lang=En&n=4A577BB9-1.
9 These reserves include used goods that are not immediately available.

capita [UNE 11]. The 10 billion mobile phones sold worldwide up to 2010 contain 2,500 t of silver, 240 t of gold, 90 t of palladium, 90,000 t of copper and 38,000 t of cobalt. A ton of mobile phones contains 10 to 30 times more gold than a ton of mined gold. In 2007, reserves of rare earth metals in manufactured products were estimated to be 448,000 t, of which 144,000 t was cerium and 137,000 t neodymium. Metallic neodymium, which is an important component of permanent magnets, can be found in computers (40,000 t), audio systems (31,000 t), wind turbines (18,000 t) and cars (18,000 t). However, anthropogenic raw material reserves are scattered and have limited potential for exploitation. For example, Nokia estimates that only 2-3% of mobile phones sold are recycled[10]. According to the European Electronics Recycling Association, only 25 to 35% of electronic boards are disassembled before shredding electrical and electronic waste in Europe. At present, high levels (>30%) of collection and recycling of end-of-life products are only achieved for certain base metals such as iron, aluminum, copper, molybdenum and the precious metals. For special metals such as rare earth elements, indium, tellurium or tantalum, global recycling rates remain low and are globally below 1% [UNE 13b]. However, the figures vary according to the regions of the world: the International Copper Study Group (2013) [11] announced, for example, recycling rates of around 40% for copper in Europe while USGS data indicates 33% in the United States for 2010-2014[12]. Some of the differences can be explained by different definitions of the recycling rate. Glöser et al. [GLÖ 13] report six different indicators of copper recycling for the period 2000-2010: the collection rate of end-of-life products (CR) and recycling efficiency rate (RER) are about 60%, the recycling rate of copper from end-of-life products (EOL-RR) is 45%, the Old Scrap Ratio (OSR) is 50%, the rate of recycled metal is 35%, and the proportion of

10 https://www.cnet.com/news/nokia-kiosks-collect-phones-for-recycling.
11 http://www.icsg.org/index.php/component/jdownloads/finish/170/1188.
12 https://minerals.usgs.gov/minerals/pubs/commodity/recycle/myb1-2014-recyc.pdf.

recycled metal to total production is 15-20%. Grädel *et al.* [GRÄ 11a] and UNEP [UNE 11] indicate that the global copper recycling rate (EOL-RR), defined as the ratio between the flow of metal actually recycled and the flow of metal in end-of-life products, is >50%. Two OSR values are reported in the UNEP (2013) report: that of Goonan [GOO 10] is 24%, and that of Grädel *et al.* [GRÄ 04] is 78%. There are therefore significant differences, even for well-documented metals such as copper. Precise quantification remains difficult due to the challenges discussed by the authors, which include measurement of flows in a large number of countries and for very different industrial sectors. Two other challenges are worth discussing:

– the recycling rates reported in the literature sometimes combine old waste with the recycling of waste production (new waste). In the first case, it is the real recycling of material in end-of-life products. In the second case, it is a matter of recycling metal that has never been used. For this reason, a high recycling rate does not necessarily mean that the value chain is good and the waste is recycled after use. A high level of recycling can also indicate poor efficiency of metal use with high scrap production during products manufacturing. In this case, the recycling stream comes from metal that has become waste before being used. The amount of energy to produce a mass of metal used is then larger than the ideal one, which corresponds to a single production cycle, if the scrap during the production of manufactured products was zero;

– some end-of-life products are recycled outside the country where they have been used, particularly in countries where the future of waste is not necessarily quantified. European regulation restricts the possibility of landfilling waste and imposes environmentally-friendly recycling conditions (which are thus expensive since they require significant investment). A consequence of these regulations is the massive export of waste to Asian countries and Africa, where recycling can take place under conditions that are less costly and not always controlled. Exports of iron, steel, Ni, Al and Cu waste thus doubled between 1999 and 2011 and exports of precious metals increased threefold[13]. Obviously, export statistics are based on those reported, and in 2005 an inspection of 18 European ports showed that half of the waste exports were illegal. This concerns electronic waste, classified as hazardous waste. The situation is similar in the United States, which has not signed the Basel Convention, where 50 to 80% of the collected waste is not recycled on the spot but exported to China. India and Asia import between 1 and 3 Mt of copper per year as waste, while annual world primary production is around 16 Mt and the global flow of copper in the form of waste is approximately 10 Mt [GLÖ 13]. Apart from the possible environmental and health consequences of uncontrolled recycling, the mass export of metals leads to other adverse effects as it prevents the significant investment in recycling and building of the infrastructure that allows us to recycle our own waste in the future.

Another limitation of recycling is the quality of recycled metals. While the recycling of pure metals such as copper does not alter the quality, and use of the metal, it is not the same for alloys, whose properties often degrade during recycling. This is the case for special steels that are recycled to produce low quality steels, or concrete and composites from wind turbine blades for example. Aluminum from aeronautics is another example[14]: traditionally, dismantling sites for old aircraft function in the same way as car scrapyards, the wrecks are dismantled to recover the reusable equipment and then the carcasses are directed to recycling facilities after rough sorting. However, an aircraft wing is made of different aluminum alloys, which are not of the same value and recycling without distinction represents a significant loss of value. A specialist subsidiary of the Suez group has teamed

13 Movements of waste across the EU's internal and external borders, EEA report no. 7/201.

14 https://www.industrie-techno.com/tarmac-aerosave-recupere-l-aluminium-pour-airbus.18129.

up with EADS and Rio Tinto Alcan to optimize the sorting of aluminum alloys after the complete mapping of their distribution on the fuselages and aircraft wings. Once this is done, the cut is made according to the metal price at the London Metal Exchange, which determines the purchase price of the aircraft carcass. When the price of aluminum is low, the cutting is rough and suitable for recasting at a refiner supplying the packaging manufacturers. On the contrary, if the price of aluminum is high, the cutting is done by following the cartography of the alloys in order to supply raw materials that are homogeneous to the Rio Tinto Alcan recasting process. The ingots produced are then rolled to produce sheets for the manufacture of new aircraft.

Obviously, what is true for large end-of-life objects that are dismantled by hand is not conceivable for miniaturized electronics, where the amount of recoverable metal is too small and the dilution is too high for manual dismantling to be cost effective. These economic aspects will be discussed in the next chapter with regard to the energy cost of recycling. However, it is already understood that if recycling is more expensive than primary production, it will remain at a low or non-existent level. An interesting example is the attempt to recycle rare earth metals from compact fluorescent lamps by the Solvay group, which opened two recycling plants in France in 2012. The process developed allowed the separation of six different rare earth metals by successions of pyrometallurgical and hydrometallurgical processes. Four years later, the factories have shut down due to lack of profitability after a sharp decline in prices of primary rare earth metals. Another problem was a shortage of raw materials with the arrival of LEDs. The market for LED lamps is evolving, but their deposits are difficult to estimate because of their theoretically very long lifetime and the lack of information on their content; the composition of LEDs is still quite confidential. However, as with all miniaturized products, complexity and dilution make the recycling of LEDs difficult; an LED lamp contains an electronic board and a printed circuit board. There are many other examples where recycling is limited by dilution and complexity, and therefore by recycling costs. Falconer [FAL 09] estimates, for example, that 60-80% of the copper contained in underwater cables linking off-shore wind farms in the United Kingdom will not be recycled for both technological and, above all, economic reasons, at current copper prices.

3

Energy Requirements of the Mining and Metallurgical Industries

Mining and metallurgy are highly energy intensive industrial activities. According to the International Energy Outlook [IEA 13], more than 20% of the energy consumed by the industry is used for steel and cement production alone. According to Nuss and Eckelman [NUS 14], the global primary energy demand for mineral raw material production was 49 exajoules (EJ) in 2008, accounting for nearly 10% of global primary energy production in the same year (520 EJ, [IEA 10a]). With industrial consumption accounting for about 40% of global consumption, about a quarter of the energy used by industry worldwide is used to produce mineral raw materials. This figure does not take into account aggregates, phosphates and potash and, in general, industrial minerals excluding carbonates.

In detail, the amount of energy required to produce a mineral raw material varies considerably depending on its nature and production processes. The cumulative energy (Ecum) to extract a mass unit of metal from a natural rock depends on many parameters, including the type of exploitation (open pit or underground mine), the hardness of the mineralized rock, the fraction of metal lost throughout the value chain, the concentration of metal sought in the rock, the degree of crushing of the ore prior to its metallurgical treatment, the metallurgical process and the nature of the mineral carrier (silicate, oxide, sulfide, carbonate, etc.). The primary energy consumed for primary production of one ton of a selection of cement and metals produced on a large scale is illustrated in Table 3.1. In the case of copper, for example, solvent extraction followed by electrolytic extraction, used for oxide ores, consumes twice (64.5 MJ per kg of refined copper produced) the amount of energy required for production of refined copper from sulfide ores by flotation followed by pyrometallurgy and electrolytic refining (33 MJ per kg of refined copper produced). Table 3.1 also illustrates the difference between primary and secondary (consumed) energy. The primary energy required for the production of primary aluminum by the Bayer-Hall Héroultst process is estimated to be about 120 MJ/kg using hydroelectricity with a 90% yield, while it is 210 MJ/kg if electricity is produced with coal with a yield of 35%, and 150 MJ/kg if produced with natural gas with a yield of 54% [NOR 07]. For this reason, the production of aluminum from ore is traditionally located near watercourses whose energy is easily transformed into low-cost electricity. More recently, aluminum production has taken place in Iceland, which has no aluminum reserves but a large amount of cheap renewable geothermal energy. Finally, Table 3.1 indicates that, recycling of non-alloyed metals after collection and separation is much more energy efficient than primary production, because it only includes the part needed for melting. Table 3.2 shows average energy consumption for the production of a larger set of metals, some of which are used in advanced technology. These data may be limited in scope because they relate to few deposits. For rare-earth metals, the data refer to the Bayan Obo reserve (China), which accounts for more than 90% of the world production of light rare-earth metals (Ca, Ce, Eu, Gd, Nd, Pr).

	Primary energy (MJ/kg) Primary production	Primary energy (MJ/kg) Recycling after collection and sorting	Consumed energy (MJ/kg) Primary production	Consumed energy (MJ/kg) Recycling
Cement	5.6[1]; 3-4[3]	-	3.1[4], 4[6]	-
Steel(*)	22.7[1]; 21.3[5]; 23.1[3]; 29.2[8]	9.7[1]; 9.84[4];14.4[8]	25[4,6]	4.5[4] (EU27)
Aluminum	211.5[1]; 194[3]; 120 (hydroelectricity)-211 (coal)[7]	10.1-17.5[1]; 8[3] (new scrap)-24[3] (old scrap)	83[4]; 93[6]	5[4]
Copper	33[1](Pyro); 64.5[1](Hydro); 45[3] (PGM, Noril'sk Mine) – 88[3] (Au-Ag ores)	5-50[1]; 1.8[3] (electronic scrap) - 28.1[3] (scrap)		
Nickel	113.5[1](Pyro); 194[1](Hydro); 150[3] (PGM, Noril'sk Mine) - 187[3] (Ni suplhides co-mined with Cu)	12.9[1]; 6.2[3]		
Zinc	35.8[1](ISP); 48.4[1](elect.); 33[3] (Au-Ag ores)-54.4[3] (Lead/Zinc ores)	4-22[1]; 18[9]		
Lead	19.6[1](BF); 32.5[1](ISP); 28.9[3] (Au-Ag ores) - 27.9[3] (Lead/Zinc ores)	9.4-11.2[1]; 0.4[3] (electronic scrap)-11.9[3] (scrap)		

Table 3.1. *Energy consumed to produce one ton of mineral raw material. 1) Rankin [RAN 11], 2) Ecoinvent 2.2, 3) Nuss and Eckelman [NUS 14], 4) Birat et al. [BIR 14], 5) IEA, 6) Gutowski et al. (2013), 7) Norgate et al. [NOR 07] for electricity with 90% efficiency, 8) Das and Kandpal [DAS 97], 9) Grimes et al. [GRI 08]. (*) For steel, primary production values are given for an initial material*

The average values in Table 3.2 also do not detail the differences in production energy of the alloys, although there are notable differences. This is illustrated by the case of steel: Fujii *et al.* [FUJ 05] report values of 20 MJ/kg for primary carbon steel produced in blast furnaces and 10 MJ/kg for the same recycled steel produced by electric arc furnace, between 40 and 72.1 MJ/kg for various stainless steels and a variation from 73 MJ/kg (100% primary) to 23 MJ/kg (100% recycled) for 304 2B steel. An accurate estimate of the production energy thus implies detailed knowledge of the composition of steel, but also of its origin and mode of production. What is true for the most produced metal in the world is obviously true for other metals.

Metal	Energy (MJ/kg)	References		Metal	Energy (MJ/kg)	References
Al	190-230	1,3, Tab. 2		Mg	270-350	2,10
Cd	17	2		Mn	52-59	2,9
Ce (oxide)	56	8		Hg	90-150	2,9,12
Cr	70-83	9,12		Mo	15-380	2,9,12
Co	130	2		Nd (oxide)	358	8
Cu	30-90	4, Tab. 2		Ni	110-200	Tab. 2
Dy (oxide)	4154	8		Pd	180000	2
Eb (oxide)	328	8		Pr	442	8
Eu (oxide)	97	8		Rh	560000	2
Ferrochrome	37	2		Sm (oxide)	9159	8
Ferronickel	160	2		Si	1000-1500	2
Gd (oxide)	320	8		Ag	1500	2
Ga	3000	2		Ta	4400	2
Au	310000	2		Te	160	2
In	2600	2		Tb	7456	8
Acier	20-30	5,2,10,Tab. 2		Sn	250	2,9
Acier inox.	304	2,6		Ti	360-750	9,13
La (oxide)	61	8		V	3700	9
Pb	20-25	2,9,11		Y	213	8
Pr (oxide)	31,63	8		Yterbium	294	8
Li	380-850	2		Zn	33-55	7, Tab. 2

Table 3.2. *Average energy consumption required for the production of mineral raw materials.1) European Aluminium Association: http//www.alueurope.eu/, 2) Ecoinvent 2.2 : http://www.ecoinvent.org, 3) http//veb.rth.edu/2.813/www/ readings/ICEv2.pdf.old, 4) Kupfer Institut: http://kupferinstitut.de, 5) World Steel Association: http://www.wolrdsteel.org/, 6) International Stainless Forum: http://www.worldstainless.org, 7) International Zinc Association: http://wwwzing.org/, 8) Koltun and Tarumajah [KOL 10], 9) Bath ICE v2.02, 10) Ashby [ASH 09], 11) Gabi software: http://gabi-software.com, 12) Rankin [RAN 11, RAN 12], 13) Norgate et al. [NOR 10a]*

Several studies have shown that Ecum is a power function of the dilution, which is the inverse of concentration (1/Cmetal) of the metal mined from the ore [GUT 12, JOH 07]. This trend is consistent with the cumulative primary production energy reported in Table 3.2, which varies between about 20 MJ/kg for iron/steel and >150,000 MJ/kg for gold and platinum. The decline in concentration of deposits exploited over time (Figures 3.9 and 3.10) would therefore increase Ecum. This situation is well documented for copper [CHA 74, NOR 10a], uranium and gold [MUD 07].

The increase in Ecum with dilution has so far been compensated by improving the energy efficiency of the techniques and means of production [GUT 12, YEL 10]. An important question is whether the same trend will continue in the future. Indeed, production costs depend on Ecum and the reserves, i.e. the economically exploitable part of the known resources, depend on the production costs. Assessing the future development of Ecum is therefore a key point for estimating not only the energy cost but also the evolution of mineral reserves. This evaluation requires an adequate formalism, which we try to establish in the next section. Other approaches based on exergy (entropy production) are proposed in the literature. These are complex approaches that will not be discussed here, but the interested reader will find valuable information in the research by Goessling-Reisemann [GOE 08] or Valero *et al.* [VAL 13] and Valero and Valero [VAL 15].

3.1. Modeling of metal production energy

3.1.1. *Primary production*

For diluted metals such as gold and platinum, for example, the cumulative production energy (Ecum) is dominated by the mechanical crushing, extraction and separation processes, whereas at high concentration (iron and aluminum for example), it is dominated by chemical separation–reduction processes. The cumulative production energy can therefore be broken down into two terms corresponding to 1) contributions of the metallurgical processes to transform the ore into metal (e1), and 2) other contributions further upstream, which include extraction mining, grinding and initial purification (e2).

The minimal value of e1 (e1,min) is equal to the free energy of formation of the mineral bearing the metal from its constituents: $e1,min = -\Delta G°f$ (J/Mol). The transformation energy of the metal-bearing minerals into pure metal (metallurgy) therefore increases with increasing thermodynamic stability of the bearing minerals. The free energy of formation is high for aluminum oxide ($\Delta G°f_{Al2O3}$ = -29.5 MJ/kg), which makes its purification much more energy-intensive than that of iron oxide hematite ($\Delta G°f_{Fe2O3}$ = -6.7 MJ/kg). This explains why Ecum (Al) > Ecum(Fe), while the concentration of these metals in the exploited deposits is of the same order of magnitude. In the case of native metals, the free energy of formation is zero. To express e1,min in MJ/kg of metal produced, conventionally used in the literature, it is necessary to divide -$\Delta G°f$, usually expressed in kJ/mol, by the molar mass of the metal to be extracted (Mmetal in grams) and the number (n) of metal atoms per formula unit of the carrier mineral:

$$\text{e1,min (kJ.g}^{-1} \text{ or MJ.kg}^{-1}) = -\Delta G°f/(M\text{metal.n}) \qquad [3.1]$$

The order of magnitude of e2 is more difficult to estimate because e2 depends on the nature of the rocks exploited. In general (see Burford and Niva [BUR 08], de Bakker [DEB 13], Silva *et al.* [SIL 12]) it is observed that 1) the grinding size depends on the size of the metal-bearing phases and their concentration, 2) the recovery rate of the metal increases with the concentration of the deposits and the intensity of the grinding in a non-linear manner, and 3) less concentrated deposits must generally be ground finer than more concentrated deposits. Different relationships between particle size and e2 are proposed in the literature for a particle size range before and after grinding [SIL 12]. All these equations correspond to an integration of the following relation, linking grinding energy (Eb) and particle size (s):

$$Eb = K.s^{-u} \ (0.5 < u < 1) \qquad [3.2]$$

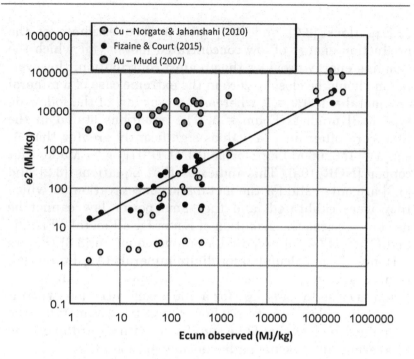

Figure 3.1. *Production energy of different metals calculated with the power laws of Norgate and Jahanshasi [NOR 10a] for copper, Mudd [MUD 07] for gold, and Fizaine and Court [FIZ 15] for a set of metals, as a function of the production energy observed. For a color version of this figure, see www.iste.co.uk/vidal/energy.zip*

Equation [3.2] indicates that the grinding energy increases as a power law of the inverse of the particle size. It tends towards infinity when the size tends towards 0. A comparable trend is observed between Ecum and the inverse of the metal concentration (1/Cmetal). Different power laws have also been proposed in the literature (Norgate and Jahanshahi [NOR 10a], Mudd [MUD 07], Gutowski *et al.* [GUT 12], Johnson *et al.* [JOH 07]), in the general form:

$$Ecum \ (MJ.kg^{-1}) = a.(1/Cmetal)^{-u} \qquad [3.3]$$

Equations [3.2] and [3.3] give a good estimate of the production energy of low concentrated metals, for which the grinding energy provides the greatest contribution. However, Ecum should be close to zero in the extreme case of a mineral or Cmetal = 100% = 1, whereas a power law of the following type [3.3] predicts Ecum = a. The equations stated in the literature often have 'a' values significantly greater than 1, e.g. a = 156266 MJ.kg^{-1} for gold [MUD 07] or 77 MJ.kg^{-1} for copper [NOR 10a]. This indicates that equations [3.2] and [3.3] are only valid for the metal and concentration for which they were calibrated, and the same power law cannot be used to describe the evolution of Ecum for all metals. Fizaine and Court [FIZ 15] proposed a unique equation [3.3] (Figure 3.1), but Ecum calculated for dilute minerals (Au, Pt and Pd) is approximately 8 times lower than observed values and Ecum is close to 8 MJ.kg^{-1} for a 100% concentration, when it should be close to zero. For concentrated metals, the contribution of e1 should be significant with regard to e2 and this term cannot be negligible (as in equation [3.3]).

Some studies propose that the separation energy of a metal from a rock is analogous to the work required to extract a pure gas from a gaseous mixture (see for example Gutowski [GUT 08]). In this case, assuming there is a perfect mixture between metal and the homogeneous matrix, the minimum work (Ws,min) required to separate one mole of rock into two fractions (metal and matrix) is proportional to the entropy of mixture (ΔSmix) :

$$Ws,min = -T.\Delta Smix = -R.T.(Xmetal.ln(Xmetal) + (1 - Xmetal).ln(1-Xmetal)) \qquad [3.4]$$

where R = 8.314 J.K^{-1}.mol^{-1} is the ideal gas constant, T the temperature in Kelvin and Xmetal (0 < Xmetal < 1) is the molar concentration of metal in the rock. For T = 298.15 K (25°C), this equation indicates a symmetrical evolution of Ws,min between the maximum value of 1.7 kJ.mol^{-1} at

Xmetal= 0.5 and 0 kJ.mol^{-1} at Xmetal = 0 and 1. The minimum work required to extract one mole of metal is given by:

$$Wsm,min = -R.T.(\ln(Xmetal) + (1-Xmetal)/Xmetal.\ln(1-Xmetal)) \qquad [3.5]$$

i.e. Wsm,min $\approx$ -R.T.ln(1/Xmetal) for diluted metals. Equation [3.5] increases with dilution and the minimum work tends towards infinity when Xmetal > 0. This trend is consistent with observations, except that in equation [5.5], Ws,min is proportional to the *logarithm* of dilution, whereas equation [3.3] indicates a power law of Ecum with dilution. In fact, the analogy between a metal in a rock and a gas mixture is clearly incomplete since equations [4.4] and [5.5] are only valid in the case where there is no interaction between the components to be separated. This situation is approached in the case of gold particles in a sand placer deposit, but certainly not for the same particles in a gangue of quartz, where grinding is necessary to free the particles prior to separation. To account for the grinding energy, it is necessary to maintain the dependence of e2 on the dilution (equation [3.3]). Two solutions are possible, the most natural being adding a term to equation [5.5] comparable to that described by equation [3.3]:

$$e2_1 \text{ (J/mol metal)} = Ws,min + a.(1/Cmetal)^{-u} \qquad [3.6]$$

This solution has the advantage of splitting e2 into the sum of a contribution corresponding to the separation and the contribution corresponding to grinding. In this case, the term 'a' must be adjusted in order to reproduce the observed values. Another approach is to consider that term 'a' of equation [3.3] is proportional to the separation energy Ws,min. The power laws linking energy and dilution proposed in the literature integrate both the grinding and separation contribution. Moreover, these laws indicate that the term 'a' increases with dilution, like Ws,min. Energy e2 for this second approach takes the following form:

$$e2_2 \text{ (J/mol metal)} = Ws,min/Cmetal^{-u} \qquad [3.7]$$

The main advantage of equation [3.7] is to ensure that e2 = 0 for Xmetal = 1, which is not the case for [3.6].

Equations [3.5] to [3.7] refer to molar concentrations, whereas metal concentrations reported in the literature refer to mass concentrations. If the metal is in its native form, the relationship between Xmetal and Cmetal is the following:

$$\text{Xmetal} = \frac{\text{Cmetal}}{c.\text{Cmetal} + d} \text{ with } c = \frac{M\text{matrix} - M\text{metal}}{100} \text{ and } d = \frac{M\text{metal}}{M\text{matrix}}$$
$$[3.8]$$

where M = molar mass (g.mol^{-1}).

In the case of complex metals in the form of oxides or sulfides for example, the metal to be extracted is contained in a mineral-bearing phase and it is necessary to replace Xmetal with Xcarrier, and Cmetal with the mass concentration of this carrier (Ccarrier = Cmetal.Mcarrier/(Mmetal.n), with n = number of metal atoms per unit formula of carrier). Finally, the overall extraction energy for e2$_1$ is the following:

$$\text{Ecum 1} \left(\frac{MJ}{Kg \text{ metal}}\right) = \alpha1. e1, min + \beta1. \left(\frac{Ws,min}{M\text{metal}.1000} + a. \left(\frac{1}{C\text{metal}^u}\right)\right)$$
$$[3.9]$$

and for e2$_2$:

$$\text{Ecum 2} \left(\frac{MJ}{Kg \text{ metal}}\right) = \alpha2. e1, min + \beta2. \left(\frac{Ws,min}{M\text{metal}.1000} . \left(\frac{1}{C\text{metal}^u}\right)\right)$$
$$[3.10]$$

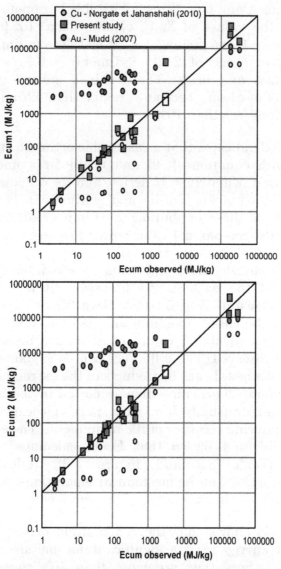

Figure 3.2. *Ecum (blue squares) calculated with equation [3.9] left [3.10] right as a function of the observed values. The empty squares represent Ecum values for Ga and In obtained with the second values of β1 and β2 listed in Table 3.3. The gray and yellow circles show the values obtained with the power laws of Mudd [MUD 07] for gold and Norgate and Jahanshahi [NOR 10a] for copper, respectively. For a color version of this figure, see www.iste.co.uk/vidal/energy.zip*

The terms α and β (> 1) are adjustable parameters. Table 3.3 shows the values of X, C, e1,min, e2, $\alpha 1, \alpha 2$, $\beta 1$ and $\beta 2$, Ws,min, Ecum1 and Ecum2 for 20 metals covering a mass concentration range of 2.10^{-6} < Cmetal < 0.5 as well as cement, glass and brick. These values indicate that the contribution of e1,min becomes greater than e2$_1$ and e2$_2$ for concentrations less than 10%.

Figure 3.2 indicates that a reasonable estimate of Ecum is possible with equations [3.9] and [3.10] for a and u $\approx$ 1, assuming that Mmatrix $\approx M$carrier, and for constant values of α and β (except for indium and gallium). In both cases, accounting for the contribution of e1,min greatly improves the estimates compared to the power law used by Fizaine and Court [FIZ 15]. There may still be significant differences between Ecum calculated and Ecum observed, but given the dispersion of the data in the literature, it is difficult to analyze the results any further. Nevertheless, it is worth mentioning the cases of indium and gallium. For these two elements, the calculated values of Ecum are much greater than those observed. The difference can be explained by the fact that both metals are byproducts of the extraction of zinc and aluminum. The Ecum values reported in the literature are therefore unusually low, since part of the grinding is carried out during recovery of the main metal. The estimated values of $\beta 1$ and $\beta 2$ so that Ecum calculated = Ecum observed (Table 3.3 and Figure 3.3), indicate that approximately 80% of the mechanical separation energy e2 is produced during the extraction of the main metal.

Figure 3.3 shows that the contribution of the chemical separation energy (e1) is significant for metals extracted from rocks where they are more than 10% concentrated, whereas the mechanical contribution e2 becomes significant below this value. The minimum value of e2 obtained with equation [3.9] for C = 100% is 1MJ.kg-1 (orange curve). Conversely, the condition e2 = 0 for C = 100% is respected with equation [3.10].

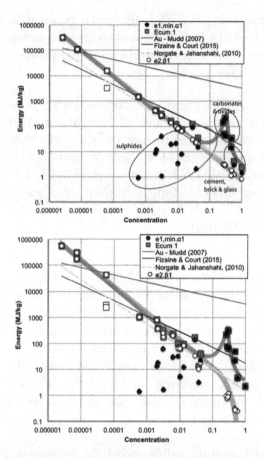

Figure 3.3. *Ecum values (blue squares and thick lines) of the chemical separation energy (e1,min black dots) and the mechanical separation energy (e2, white dots and orange curve) calculated with equations [3.9] top and [3.10] bottom. The Ecum values obtained with the power laws of Mudd [MUD 07] for gold, Norgate and Jahanshahi [NOR 10a] for copper and Fizaine and Court [FIZ 15] are also reported. The empty squares represent the Ecum values for Ga and In obtained with the second values of β1 and β2 listed in Table 3.3. For a color version of this figure, see www.iste.co.uk/vidal/energy.zip*

Even if Figure 3.3 seems to indicate the contrary, the term e1,min is not dependant on concentration, it only depends on the nature of the carrier mineral. The least concentrated metals (Pt and Au) may be in their native form in natural reserves*1* (-$\Delta G°f$ = 0), more concentrated metals (Fe, Al, Mn, Mg) are in the form of oxides or carbonates (high -$\Delta G°f$) and metals of intermediate concentrations are often in the form of sulfides.

The yellow line in Figure 3.3 shows the variation in Ecum obtained with the power law of Norgate and Jahanshahi, [NOR 10a] for copper calibrated over a concentration range between 0.5 and 5% (ENJ = 77.585.C(%)-0.857). Over this concentration range, ENJ is almost parallel to e2 and Ecum, the slope in our case is slightly higher since we assume that u = 1 when u = 0.857 in the equation of Norgate and Jahanshahi [NOR 10a].

The equation calibrated using 34 metals by Fizaine and Court [FIZ 15] (EFC = 279.25.C(%)-0.60026) has a much lower slope. This equation underestimates the production energy of less concentrated metals such as gold, platinum or palladium and that of concentrated metals such as iron, aluminum, manganese or magnesium. It overestimates the production energy of metals of intermediate concentrations such as copper, mercury or lead or those of highly concentrated resources such as cement or brick (see Table 3.3).

Finally, the Ecum values obtained using the calibrated power law for gold by Mudd [MUD 07] (EM = 156266.C-0.2793) differ greatly from the values observed for all metals, except for those whose concentration is similar to that of gold (Pd, Pt). The lower value of the exponent "u" for gold in equation EM than for copper (ENJ) is consistent with the decrease observed for the same exponent in the relationship between grinding energy and particle size (equation [3.2], see [SIL 12]). This observed decrease is not consistent with our assumption that u = 1 in equation [3.11], but considering the

1 Platinum was historically exploited in its native form, platinum arsenides are now the main source of this metal.

data we have and the results reported in Figures 3.1 to 3.3, our assumption seems reasonable.

Equations [3.9] and [3.10] can now be used to calculate the evolution of Ecum with the decrease in reserve content. Different examples are shown in Figures 3.4 and 3.5, which show that the log-normal evolution of Ecum with the mass concentration is consistent with the trends published in the case of copper and nickel (Figure 3.4) or gold (Figure 3.5). We find that the energy values that we predicted for copper and nickel are lower than those reported in 1974 by Chapman [CHA 74] for reserves with C > 1% (red arrow in Figure 3.4). This can be explained by the advances in technologies that have allowed a considerable reduction in energy consumption by the mining industry and mineralogical transformation over 40 years. For this reason, the Ecum for copper produced from current reserves (50 MJ.kg^{-1} for C = 1.3%) is identical to the value of Chapman [CHA 74] for 3% deposits.

To obtain an Ecum function that can represent both values from 1974 and current values, we must integrate in equations [3.9] and [3.10] a dependence of α and β with time to account for the improvement in energy efficiency. From the previous observations, we find that α and β should decease with time. The calculated and estimated values for gold shows similar features, even if the values predicted with equations [3.9] and [3.10] seem to overestimate the production energy of this metal with a concentration < 10 g.t^{-1} with regard to values reported by Mudd and Diesendorf [MUD 08]. This observation may reflect better energy efficiency of the gold production chain than that for other materials used for the calibration of α and β. The Ecoinvent[2] value used for the regression (red square in Figure 3.5), is much greater than values reported by Mudd and Diesendorf [MUD 08].

---

2 http://www.ecoinvent.org.

Mporteur			e1,min (MJ/kg)	Cmetal (g/grock) Used[a]	mean[b]	PE1976[c]	XMetal	Ws,min (MJ/kg)	1/C	α1	β1	Ecum,1 (MJ/kg)	α2	β2	Ecum,2 (MJ/kg)	Ecum (MJ/kg) Used[a]	FC2015[b]	Tab.2
1	brick	CaCO3	*0.30*	1.00E+00	1.00E+00	1.00E+00	1.000	4.24E-06	1.00E+00	4.9	0.9	2	7.5	8	2	2[e]	-	2
100	cement	CaCO3	*0.60*	5.60E-01	5.60E-01	5.60E-01	0.694	2.20E-02	1.44E+00	4.9	0.9	4	7.5	8	5	4[e]	-	4
60	glass	SiO2	*3.00*	4.66E-01	4.66E-01	4.66E-01	0.757	3.03E-02	1.32E+00	4.9	0.9	16	7.5	8	23	15[e]	-	15
158	Fe	Fe2O3	*6.75*	4.81E-01	4.81E-01	4.20E-01	0.627	1.65E-02	2.08E+00	4.9	0.9	35	7.5	8	51	23	23	20-25
102	Al	Al2O3	*29.30*	2.55E-01	2.55E-01	4.20E-01	0.559	2.98E-02	3.92E+00	4.9	0.9	147	7.5	8	221	200	212	190-230
86.93	Mn	MnO2	*8.49*	3.00E-01	3.98E-01	3.00E-01	0.438	4.46E-02	3.33E+00	4.9	0.9	44	7.5	8	65	57	57	52-59
79.86	Ti	TiO2	*18.61*	2.80E-02	9.36E-02	2.80E-02	0.057	1.19E-01	3.57E+01	4.9	0.9	122	7.5	8	174	430	430	360-750
95.545	Cu	CuO	*1.63*	1.30E-02	4.82E-02	1.30E-02	0.016	1.33E-01	7.69E+01	4.9	0.9	73	7.5	8	94	62	64	30-90
152	Cr	Cr2O3	*10.07*	2.74E-01	2.74E-01	3.30E-01	0.421	2.64E-02	3.64E+00	4.9	0.9	52	7.5	8	76	64	64	83
219.71	Sn	SnO2	*4.38*	8.60E-03	-	8.60E-03	0.006	6.99E-02	1.16E+02	4.9	0.9	120	7.5	8	98	207	207	250
264.655	Hg	HgS	*0.22*	3.20E-03	1.14E-02	3.20E-03	0.001	7.11E-02	3.13E+02	4.9	0.9	267	7.5	8	179	409	409	130
97.4	Zn	ZnS	*3.04*	4.20E-02	1.42E-02	4.20E-01	0.063	9.51E-02	2.38E+01	4.9	0.9	35	7.5	8	41	45	45	49-55
90.7	Ni	NiS	*4.04*	6.00E-03	4.64E-02	6.00E-03	0.010	1.53E-01	1.67E+02	4.9	0.9	161	7.5	8	234	190	-	180-200
279.6	Ag	Ag2S	*0.19*	5.90E-04	4.16E-03	5.90E-04	0.001	7.55E-02	1.69E+03	4.9	0.9	1442	7.5	8	1024	1582	1582	1500
192.09	Mo	MoS2	*2.35*	2.10E-03	3.03E-03	2.10E-03	0.001	9.85E-02	4.76E+02	4.9	0.9	416	7.5	8	393	148	148	15-380
79.86	Co	CoS	*1.80*	2.00E-03	-	5.80E-04	0.004	2.01E-01	5.00E+02	4.9	0.9	434	7.5	8	818	322	322	130
271.26	Pb	PbS	*0.41*	3.69E-02	3.69E-02	5.90E-02	0.016	4.70E-02	2.71E+01	4.9	0.9	25	7.5	8	13	25	25	25-50
158	Cu	Cu2S	*0.68*	1.00E-02	4.82E-02	1.30E-02	0.016	8.07E-02	1.00E+02	4.9	0.9	88	7.5	8	70	78	33	30-90
158	Cu	Cu2S	*0.68*	1.00E-02	4.82E-02	1.30E-02	0.016	8.07E-02	1.00E+02	4.9	0.9	88	7.5	8	70	78	33	30-90
84.3	Mg	MgCO3	*42.31*	2.88E-01	-	2.88E-01	0.625	3.11E-02	3.47E+00	4.9	0.9	210	7.5	8	318	437	437	27-350
183.84	MnWO4	W	*8.40*	3.10E-03	7.71E-03	3.10E-03	0.002	9.96E-02	3.23E+02	4.9	0.9	315	7.5	8	320	357	357	-
	Ga		*10*	2.00E-5	2.00E-5	-	7.14E-05	2.56E-01	2.00E+04	4.9	0.9-0.2	17010	7.5	8-0.6	41027		-	3000
	In		*10*	2.00E-5	-	-	4.39E-05	2.40E-01	2.00E+04	4.9	0.9-0.1	17010	7.5	8-0.5	38406		-	2600
106.42	Pd	Pd	*0.00*	7.59E-06	7.59E-06	0.00E+00	7.14E-06	2.99E-01	1.32E+05	4.9	0.9	111925	7.5	8	315321	180000	5500	180000
195	Pt	Pt	*0.00*	2.66E-06	2.66E-06	2.90E-06	1.37E-06	1.84E-01	3.75E+05	4.9	0.9	319129	7.5	8	553793	190000	27050 / 0	190000
197	Au	Or	*0.00*	7.49E-06	7.49E-06	6.10E-06	3.80E-06	1.70E-01	1.33E+05	4.9	0.9	113417	7.5	8	181058	310000	68400	310000

Table 3.3.

NOTES ON TABLE 3.3.– a) Values used for regressions (Figure 3.2), b) mean of the max and min for values of Johnson *et al.* [JOH 07] reported in Gutowski *et al.* [GUT 12], c) concentration in 1973 reported by Philipps and Edwards [PHI 76], d) values reported in Fizaine and Court [FIZ 15], e) Birat *et al.* [BIR 14], f) for Ga and In, the second values of ß1 and ß2 are those required to reproduce the observed values of Ecum. The concentration values used for the regression (a) are those reported in (b) if they are less than the values reported by Philipps and Edwards [PHI 76], or the values of Philipps and Edwards (1976) in the opposite case. The Ecum values used for the regression are those listed in Fizaine and Court [FIZ 15] except for Pd, Au and Pt.

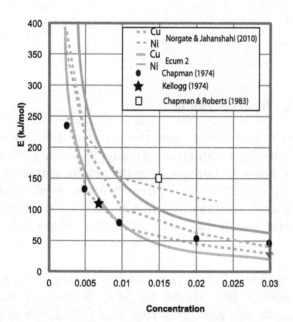

Figure 3.4. *Comparison of production energies of copper (yellow curves) and nickel (blue curves) estimated by different authors and calculated with equation 10 (Ecum2). The red arrow indicates the improvement in production energy in 40 years. For a color version of this figure, see www.iste.co.uk/vidal/energy.zip*

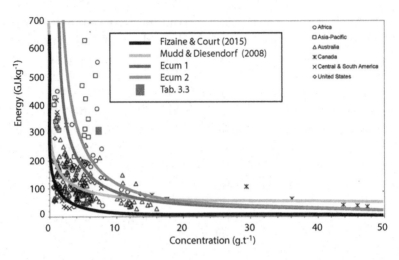

Figure 3.5. *Production energy as a function of the concentration of gold deposits (according to Mudd and Diesendorf [MUD 08]). The yellow line is calculated with the equation proposed by Mudd and Diesendorf [MUD 08], the red and green curves with equations [3.9] and [3.10] and the black curve with the equation of Fizaine and Court [FIZ 15]. For a color version of this figure, see www.iste.co.uk/vidal/energy.zip*

3.1.2. *Energy from recycling*

The recycling of metals has three major differences with regard to primary production. Firstly, consumer goods put elements in contact that are usually not in contact in their natural environment. This involves a mechanical mixture of components such as plastics with metals for example, or a chemical mixture in the case of metal alloys. Next, the variability in the type of end-of-life products to be recycled is greater than that for mined minerals. Finally, the primary production is focused on a fraction of the mined rock, the rest is not processed but stored as slag heaps. In contrast, consumer goods are separated into many fractions to ensure almost complete recycling of the initial mass. There are some similarities between primary production and metal recycling: rocks like small end-of-life products (batteries for example) are crushed and the fractions are separated before undergoing pyro- and hydro-metallurgic processing. Equations [3.9] and [3.10] derived to estimate the production energy of primary metals may be used to estimate the total energy of recycling, but it requires some adaptations:

1) Equations [3.9] and [3.10] are valid for estimating the energy required for the production of a single substance from a rock assumed to contain only two phases, the metal or the carrier phase on the one hand, and the matrix on the other. To derive an equation that integrates the energy contribution of all the phases present in the product to be recycled, one must return to the equation of mixing taking into account all its constituents, which is written for an ideal mixture:

$$Gmix = \sum_1^n Xi.\mu i = \sum_1^n Xi.(\mu°i + RTlnXi) \qquad [3.11]$$

with X: molar fraction of phase i in the mixture consisting of n phases, μ: chemical potential, $\mu°$: chemical potential of the pure phase (X = 1). The minimum separation work (Ws,min) is:

$$Ws, min = -RT.\sum_1^n Xi.lnXi \qquad [3.12]$$

Equation [3.4] is therefore a special case of equation [3.12] for a two-phase mixture. If the number of phases present in the end-of-life product to be recycled increases, the minimum separation work also increases. The separation energy thus increases with the complexity and the number of constituents of a manufactured product.

2) End-of-life products often contain metal alloys. Equation [3.12] represents the separation energy contribution of the different phases, which can either be pure components or alloys. The separation of the different alloy metals from the same phase requires additional energy, but for the sake of simplicity, we consider hereinafter that all the

phases are pure. In this case, the recycling energy is that necessary to bring the metal to its melting temperature plus the melting energy (approximately 0.04 MJ/kg for lead and 1 MJ/kg for copper).

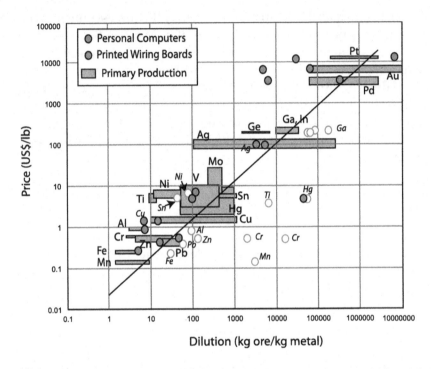

Figure 3.6. *Metals prices as a function of primary concentrations (values of Johnson et al., 2007 and Table 3.3 for In). Circles represent prices in terms of concentrations in end-of-life products (Johnson et al. [JOH 07]). Colored circles: recycled elements, open circles: not recycled. For a color version of this figure, see www.iste.co.uk/vidal/energy.zip*

3) As for natural rocks, the grinding energy of end-of-life products varies in s^{-u} with s the grinding size. However, two major differences emerge for recycling:

i) an additional term must be added to take into account the shape ratio of the elements to be separated, the assumption of spherical grains to be separated in the case of natural rocks does not apply to end-of-life products [UNE 13b]. The release of the metal depends on the shape ratio of the bearing phase, but also on the hardness contrasts between the different phases to be separated;

ii) the cohesion between the phases to be separated is different and variable (bonding, welding, crimping, bolting etc). For these reasons, and given the extreme variability of end-of-life products, it is impossible to give a single equation as with equations [3.9] or [3.10] to estimate an average recycling energy of metals. However, several studies indicate that for a given end-of-life product, the recycling of metals depends on their concentration and their price. Johnson *et al.* [JOH 07] show that only metals above a straight line in a price vs dilution graph are recycled from different types of end-of-life products (personal computers, printed maps, automobiles).

Below this line, prices are too low for recycling to be profitable at the concentration of metals in the product. Johnson *et al.* [JOH 07] also showed that the limit corresponds to the price–dilution relation defined by the primary production of metals. This is illustrated in Figure 3.6, where the metal price and dilution values are plotted against the dilution data reported by Johnson *et al.* [JOH 07]. Metals in dilution equal to or less than those observed in natural deposits are recycled (colored circles), while the others (open symbols) are not recycled. The four exceptions are tin and nickel not recycled in printed maps, and mercury and platinum from laptops that are recycled.

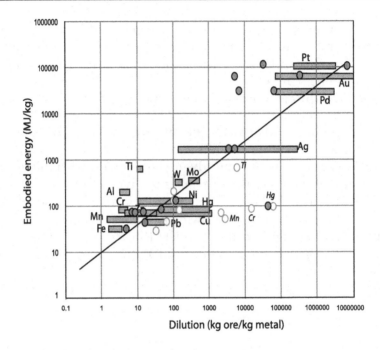

Figure 3.7. *Primary production energy of different metals according to their dilution (= 1/mass concentration). The energy values are those reported by Gutowski et al. [GUT 12]. The grey rectangles represent the dilutions in the natural deposits and the circles in end-of-life products (according to Johnson et al., 2007). Shaded points: recycled elements, empty circles: not recycled. For a color version of this figure, see www.iste.co.uk/vidal/energy.zip*

We have seen above that for primary metals concentrated < 2%, Ecum varies linearly with concentration. For this concentration range, a relationship between recyclability and energy similar to that observed between recyclability and price is expected. This is illustrated in Figure 3.7, which compares the primary production energy of metals in deposits and end-of-life products as a function of their dilution. Recycled metals are predominantly located above the energy curve versus 1/C. Since the dilutions of metals are different in printed maps and computers, their recyclability is also different. In detail, there are some inconsistencies, such as platinum in computers that is recycled despite the concentration seeming to be lower than that of primary deposits (a similar observation is made in Figure 3.6). However, Figures 3.6 and 3.7 seem to be good indicators of the recycling potential of end-of-life products. They easily explain why some low-concentration metals are not recycled while recycling is technologically feasible. They also illustrate that the recycling potential is controlled by the price versus dilution relationship of primary production.

3.2. Conclusion

The figures given in Tables 3.1 and 3.2 are the current values of energy consumed. This energy changes over time for two reasons:

– Improvements in technologies have made it possible to strongly reduce the production energy of mineral resources with time. In the case of Iron, production energy per unit mass was reduced by a factor of three in 70 years (Yellishetty *et al.* [YEL 10]). This corresponds to an improvement of approximately 2%/year. For copper, a report by the US Bureau of Mines [ROS 76] indicates an average consumption of 94.5 MJ/kg of refined copper produced in 1963, while data published by the Chilean Copper Commission for the production in 2013 indicate an average consumption of 57 MJ/kg [COC 14]. We have also seen previously that Ecum calculated for the current conditions of concentration and grinding size is equal to that reported in 1974 for deposits two to three times as concentrated.

– The progressive decrease in metal content of the deposits (Figures 3.9 and 3.10). As discussed in the previous chapter, the gradual decline of ore grade causes an increase in energy demand and extraction costs, which are inversely proportional to the dilution. It is therefore possible to evaluate the evolution of the energy demand if it is possible to estimate the future evolution of the average concentrations over time. A simple approach is to use the observed relationships between the cumulative tonnage of metal produced and the concentration (see Gerst [GER 08]; Vieira *et al.* [VIE 12] for copper). The variation in concentration over time can be estimated from the observed cumulative tonnage–concentration changes (Figure 6.12) if we make the extreme assumptions that i) the global production of copper is 60% supplied by the same copper porphyries since 1900, ii) the reserves increase as cumulative production, and iii) the increase is solely due to the decreasing concentration of exploited deposits. The evolution of the energy of production over time at constant technology is shown by the black line in Figure 3.8.

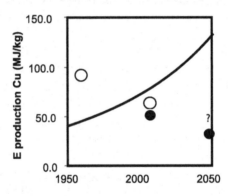

Figure 3.8. *Evolution of energy consumption for the production of primary copper from copper porphyries depending on a concentration decrease over time and with the influence of current technologies. The empty circles indicate the actual values observed in 1963 and 2013. The black circles show the mean global values for all types of deposits combined*

This energy should increase exponentially over time. Yet historical data indicates that instead of increasing, production energy has declined in recent decades. This means that, at least in the case of copper, the decline in concentration of exploited deposits has been largely offset by the use of increasingly efficient production technologies, which have reduced or maintained the extraction energy and costs. Maintaining these costs with lower concentrations has increased reserves (the amount of metal available under viable economic conditions) over time, as the low-concentrated deposits are more abundant than highly concentrated deposits. A similar decrease in production energy is observed for steel and aluminium [GUT 12, YEL 10]. Can this trend continue in the future? If an improvement in the energy efficiency of primary production is still possible, it becomes increasingly difficult when Ecum approaches the thermodynamic limit. For steel, this limit is about 10 MJ/kg. With the optimistic assumption that the energy consumption of primary production will continue to decline at the same rate as observed since the 1950s (-2%/yr), the thermodynamic limit would be reached in 2060. As the second law of thermodynamics tells us that this limit can only be approached, it is therefore likely that improved energy efficiency in the case of steel and iron will continue at a slower pace than in the past. Gutowski *et al.* [GUT 12] estimate that the future reduction in global energy used for the production of mineral resources, including recycling, will not exceed 55% of the current value. Improvements in technology and economic conditions will fix the rules, as they control the production costs and the profit level of the mining industry (see section 6.2). If energy prices and production costs allow the mining industry to exploit lower ore grade deposits, reserves will continue to increase over time. It will be possible to exploit new resources that are less accessible (deep or at sea, or even extraterrestrial). The energy required for recycling is not easy to estimate because it depends on many factors including the concentration of metals in end-of-life products, the complexity of the products to be recycled, and the decommissioning technologies. These parameters are extremely variable depending on the product and metal recycling. Technological advances could decrease energy and financial costs, but the complexity and miniaturization of products and the dilution of metals complicate the situation.

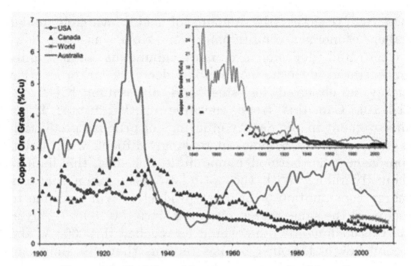

Figure 3.9. *Evolution of the copper ore grade of deposits
exploited between 1900 and 2010 (Insert: between 1840 and 2010).
Source: Mudd [MUD 13a], Mudd et al. [MUD 13b]. For a color
version of this figure, see www.iste.co.uk/vidal/energy.zip*

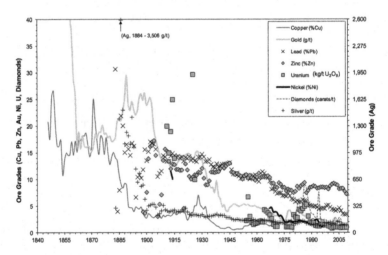

Figure 3.10. *Decrease of ore grade of various Australian deposits over
time, including copper (red curve). Source: Mudd [MUD 10]. For a color
version of this figure, see www.iste.co.uk/vidal/energy.zip*

Raw Materials for Energy

4.1. Overview of needs by major areas of application in the energy sector

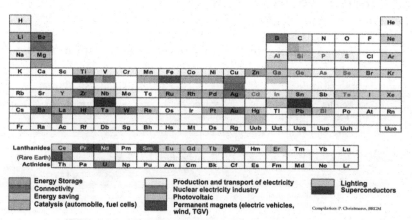

Figure 4.1. *Elements with a mineral origin (except rare gases Ne, Ar, Kr, Xe) used in technologies in the energy sector. Source: Patrice Christmann, Rapport Ancre (2015)[1]. For a color version of this figure, see www.iste.co.uk/vidal/energy.zip*

1 http://www.allianceenergie.fr/imageProvider.aspx?private_resource=1297&fn=An cre_Rapport_2015-Ressources_minerales_et_energie_0.pdf.

The raw material requirements for the energy sector are varied. These include energy production, its storage, its distribution, the transformation of fossil hydrocarbons into fuel, and also energy-saving technologies. Energy-saving technologies are also varied, such as for instance the use of lighter and more resistant alloys in transport, low-energy lamps, or technologies requiring special alloys for extreme conditions, such as in the case of nuclear reactors or reactor turbines for instance. An overview of the main energy sectors concerned by the use of each element of the periodic table is given in Figure 4.1 and the most common uses are developed hereafter, greatly inspired by the work of [TAL 11] and [ZEP 14].

4.1.1. *Hydrocarbons, coal and catalysis*

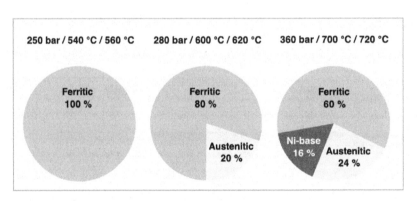

Figure 4.2. *Composition of steel for sub, super and ultra-supercritical pulverized coal boilers (ALSTOM). For a color version of this figure, see www.iste.co.uk/vidal/energy.zip*

The exploration and production of oil and gas are totally dependent on specialized steels. The acquisition of seismic data on land or sea, drilling and oil pipelines require different alloys of steels with chromium, nickel, molybdenum, manganese, cobalt, vanadium or tungsten. The extraction of highly buried hydrocarbons up to several thousand meters deep under very corrosive conditions is based on the availability of steels suitable for piping and that must withstand high pressures and temperature differences. Gas (40% of conventional reserves are acidic) is processed, shipped and stored in steel tanks. Large quantities of low grade steel are used for mine tunnels and more specialized steels for coal mining machinery. Coal-based power generation chains are evolving, in particular towards higher-yielding technologies, obtained by processes operating at higher temperatures and pressures, and requiring the use of special steels (Figure 4.2). Electricity generation requires another group of specialized steels in the manufacture of turbine blades. Solvents used for CO_2 capture are often corrosive amines, which also require special steels containing elements such as Cr, Co, Mn, Mo, Ni, Nb and Hf, Re, Ta, Y. Many other metals are required in the mining of hydrocarbons and coal, such as copper, which is an essential component of modern mining engines.

Approximately 95% of new cars are equipped with catalytic exhaust systems to reduce CO and NOx emissions. They contain combinations of platinum-group metals (PGM: Pt-Rh-Pd) and rare-earth oxides [BOR 01]. On average, the catalytic converters contain between 2 g [YEN 07] and 10 g [CUL 11] of Pt-Rh or Pt-Pd and according to the equipment and level of hybridization, between 100 g and 7,000 g of rare elements. There are no PGM substitutes in catalytic converters, which makes their recycling crucial. The USGS estimates that 155,000 kilograms of PGM were recycled worldwide in 2013 as waste. Catalysts are also used to improve the production of light hydrocarbons during the process of cracking (FCC process) heavy oils and to reform the residual heavy fractions after crude distillation. The catalysts used are zeolite-type synthetic minerals containing PGMs (mainly rhenium) and rare-earth elements (La and Ce) [RAH 11]. Approximately 5 kg of catalysts are used per liter of heavy oil entering the FCC process. The content of rare-earth elements can range from < 1 % to 3% [BAS 06]. In

2008, 1,668 billion liters of hydrocarbons were processed by FCC [DAV 09], which required about 16,000 metric tons of rare-earth elements, including 1,600 tons of cerium and 14,500 tons of lanthanum. In 2010, 9.24 tons of rhenium were used in the petroleum industry.

4.1.2. Nuclear

This energy sector requires specialized steels and materials, including lithium and rare-earth elements to control reactions and confine reactors, indium for control rods and reactors, and tin, zirconium, niobium, iron, nickel-chromium (zircaloy) for fuel cladding. As with all electric generators, nuclear power plant turbines use special steel alloys to ensure sustained yields at high temperatures and high constraints.

4.1.3. Production of electricity from renewable energies

Steel is needed to build the towers and nacelle housing the generator and the mechanism transforming the energy of the rotating blades into induced electricity in strong magnetic fields. This magnetic field is generated either by permanent magnets rich in rare-earth elements (neodymium, praseodymium and dysprosium), or by conventional magnets and copper windings forming an armature collector. Copper is also present in the inductive and auxiliary poles of the generator. At present, the quantities of rare-earth elements used for wind turbines are around 200 kg/MW of electrical power, but the increase in the price of rare-earth elements stimulates technological innovation and research to significantly reduce their use in permanent magnets (section 4.2). Two types of wind turbines are in operation: those with a gearbox and those with direct drive, which offer greater reliability and require less maintenance, but require larger magnets and therefore more rare-earth elements.

Hydroelectric production is also dependent on concrete and steel for forced pipes, abrasion-resistant turbine blades and cavitation.

Photovoltaic systems that convert solar energy directly into electricity use two different technologies: silicon cells, which account for about 90% of the market, and multilayer technology based on combinations of elements such as gallium, arsenic, cadmium, tellurium, copper, indium and selenium. Both technologies essentially use silver as a contact material. New semiconductors are under development, e.g. organic compounds, but they are not yet commercially viable. Solar thermal technology, which is used to heat domestic water, is dominated by copper and zinc. Finally, Concentrated Solar Power (CSP) systems use solar energy to heat fluid with a higher boiling point than water, which is stored in a tank. Heat exchangers use this stored energy to produce steam that drives turbines and electricity generators. The reflectors used to concentrate solar energy have polished surfaces, usually with a coating of aluminum or silver. The scale of CSP installations is illustrated, for example, by three solar power stations in Andasol in Spain, each with 200,000 parabolic mirrors and capable of generating 50 MW of electricity.

4.1.4. *Electricity transmission and grids*

National and transnational electricity grids have been constructed using large quantities of aluminum and copper for cables, generators and transformers, as well as steel for towers. As outlined in a World Nuclear Association report published in 2016[2], these grids need to undergo a major transformation in order to keep up with the changing

2 http://www.world-nuclear.org/information-library/current-and-future-generation/electricity-transmission-grids.aspx.

production patterns and the integration of solar and wind systems that produce distributed and fluctuating energy.

Germany, which is replacing its electricity generation system from fossil fuels located in the south with a system integrating renewable sources located to the north on the Baltic coast, is one example. New long-distance transmission networks are also planned to connect large offshore wind farms in the North Sea to European power grids. At the European level, the Ten Year Network Development Plan envisages the installation in Europe by 2030 of approximately 45,000 km of new lines including 18,000 km of undersea DC high voltage lines (copper cables), 5,000 km of overhead lines (Fe-Al cables), 18,000 km of AC (Fe-Al) power lines and 5,000 km of old lines to be upgraded. The *Réseau et Transport d'Electricité* (RTE) estimates that 700 billion dollars of investment in the 16 largest grids representing 70% of the world's electricity and 2.2 million km of lines will be required by 2022, in part because of the integration of renewable sources.

According to the International Energy Agency, India, China and the Middle East will account for 60% of the growth in energy demand in 2035. China is developing a sophisticated network because its coal deposits are located in the north, its main wind potential to the far west and its nuclear power stations on the coast. The grid saw rapid growth of 40,000 km of ultra-high voltage (800 to 1,000 kV AC) lines in a few years, and in a report published on the National Energy Administration (NEA) website, China Electric Power News said the country was aiming to double the length of its high voltage lines between 2014 and 2020. About half of Bangladesh's 160 million people and 600 million Indians do not have access to electricity. The problem is not only the underproduction, but also the dilapidated grid whose online losses can reach 60%, and the lack of connections between regional grids[3]. According to Bloomberg, China's State Grid Corp., which is the world's largest electricity company, is studying, along with Japan's Softbank Group Corp. and South Korean and Russian partners, the opportunity to develop the grid of Northeast Asia [4]. After the Fukushima accident, Japanese officials estimated that such a network from India to Japan would ensure a stable supply of electricity with a high proportion of renewables. In 2016, the same Chinese company proposed that a global investment of some US $ 50 trillion be pooled to develop a global ultra-high voltage grid by 2050, which would link each continent and country from the poles to the equator. This possibility, taken up by the Chinese President Xi Jinping at the United Nations summit on sustainable development in September 2016, appears to be a natural evolution that would make it possible to fully integrate sustainable energy into a single grid.

Developing countries are not the only ones developing their grids. In the United States, for example, a 2011 MIT report[5] shows that the US grid faces a number of serious challenges over the next two decades. In 2009, the website Grist [6] reported the following analysis by former energy secretary Bill Richardson: "We're a major superpower with a third-world electricity grid". "The average age of equipment that makes up the US infrastructure is more than 40 years old and many components were designed and installed before the Second World War". As elsewhere, a major shift towards renewable energy sources would require an upgrading of transmission technologies and grid. Numerous and varied raw material are needed to build and connect the

---

3 https://www.siemens.com/innovation/en/home/pictures-of-the-future/energy-and-efficiency/power-transmission-hvdc-in-india.html.

4 https://www.bloomberg.com/news/articles/2016-03-30/china-state-grid-eyes-asian-super-network-with-partners-help.

5 http://energy.mit.edu/research/future-electric-grid.

6 http://grist.org/article/2009-10-13-our-old-electric-grid-is-no-match-for-our-new-green-energy-plans.

new generators, the cables, the sophisticated switching and control systems, junctions, compensation stations and the entire power infrastructure (transformers, inverters, rectifiers, etc.). Yang Changhua, senior analyst at Antaike estimated that more than 1 million tons of copper had been used in electricity transmission projects in China in 2014 alone[7]. Ben Carstein, Head of Sector Analysis at Rio Tinto Copper, estimates that over the next 12 years, the amount of copper in the Chinese power grid is expected to double and that this will require 15 million additional tons of copper[8]. This amount is equivalent to almost one year of the world's current production, but these figures depend on the technologies used. China State Grid has in fact adopted aluminum, lighter and cheaper than copper, for its long-distance transmission grid.

A first-order approximation of the raw material requirements for transmission of electricity at the global level can be made on the basis of existing electrical grids and consumption in developed countries. Ecofys-WWF [ECO 14] proposed an estimate using copper intensities of 3.8 t/km and 2 t/km for high-voltage (500,000 km) and medium-voltage (1,100,000 km) lines in Germany, which consumes 600 electric TWh per year. This translates into 6,800 t/TWh of copper, i.e. 130 Mt of copper for the 20,000 TWh/yr of electricity consumed globally in 2012 and 240 Mt for the 36,000 TWh/yr consumed in 2050 according to the Bluemap scenario of the IEA [IEA 10b]. The requirements in 2030 would be 612 Mt of copper for the scenario of Garcia-Olivarès et al. [GAR 12], who predicted a consumption of 109,000 TWh/yr globally. The requirements would therefore be 3.3 Mt/yr of copper for the Ecofys-WWF and Bluemap scenarios, and 17.5 Mt/yr for the Garcia-Olivarès scenario. These values, which represent between 20 and 100% of the current annual global production, are extremely high and

---

7 http://www.reuters.com/article/china-power-transmission-idUSL4N1171UP20150901.
8 http://m2m.riotinto.com/issue/3/article/copper-solving-societys-challenges.

overestimated. In fact, high-voltage overhead cables are primarily made of Fe-Al, and the copper intensities determined by Ecofys-WWF are not representative of the reality. Garcia-Olivarès *et al.* [GAR 12] estimate that the amount of copper required for the electricity transmission infrastructure in their scenario should be 38 Mt, a copper intensity 14 times lower than that estimated by Ecofys-WWF. Harrison et al. [HAR 10] have estimated that 10.5 Mt of raw materials were used to build the UK electricity transmission system (53% concrete, 2.4% aluminum, 5.7% steel and 0.8% copper), which transports 350 TWh/yr. These values are equivalent to a copper intensity of 210 t/TWh, 30 times less than the value estimated by Ecofys-WWF and half that estimated by Garcia-Olivarès *et al.* [GAR 12]. Extrapolating the material intensities of Harrison *et al.* [HAR 10] globally, the material requirements for the next 35 years would range from 1.17 Mt/yr (Bluemap) to 6.6 Mt/yr (Garcia-Olivarès) of steel, 7 to 40 Mt/yr of concrete, 0.6 to 3.15 Mt/yr of aluminum and 0.1 to 0.54 Mt/yr of copper. These latter values are now considerably lower than the forecasts by Antaike or Rio Tinto Copper mentioned above for (approximately 1 Mt/yr copper for China alone). As with other energy sectors, these rapid comparisons illustrate the difficulty of estimating the raw material requirements for energy transmission, which could be in the range of 1 to 2 Mt/yr copper. Added to this is the distribution from the high-voltage grid, which according to the International Copper Association currently represents up to 3 Mt/yr of copper. This value will obviously increase with installed capacity and urbanization. A copper demand of around 5 Mt/yr could be predicted for the global electricity transmission sector alone in the coming years, which would represent 25% of current world production. The values for concrete could be around 100 to 500 Mt/yr, 10 to 50 Mt/yr for steel and 8 to 15 Mt/year for aluminum (15 to 30% of the current world production). These estimates assume that future technologies will be the same as those currently used.

Nevertheless, new types of conductors are envisaged, such as iron-rare-earth alloys, or tin- or copper-silver alloys for high-voltage continuous currents.

4.1.5. *Electricity storage*

Electrical storage is essential in the field of transport (hybrid and electric vehicles) or nomadic applications (mobile electronic devices). In 2010, about 20,000 tonnes of cobalt were used for the manufacture of batteries (76,000 tonnes produced worldwide). Other metals are also used, such as nickel or manganese combined with lithium or lantane, cerium, praseodymium, neodymium and other rare-earth elements in NiMH batteries (13,000 tons of rare-earth elements were used in the manufacture of NiMH batteries in 2010 [MOR 11]). Currently, the quantities of lithium metal in Li-ion batteries are <1 g in mobile phones, approximately 10 g in computers, 3.2 kg in electric vehicles and 1.3 tonnes for 8 MWh storage. World production of lithium has doubled since 2003 to reach 35,000 t/yr. By 2015, there were 400 million smartphones and more than 200 million tablets with Li-ion batteries. To meet the needs of electric vehicles, the future increase could be much greater. According to García-Olivarès *et al.* [GAR 12]), if all current transport was converted into electric vehicles using Li-ion batteries over the next 30 years, 8 Mt of Li would be used in the transport sector only. The demand for lithium and cobalt was estimated by Ecofys-WWF [ECO 14] for 3.3 billion light vehicles in 2050 with batteries containing an average of 3.2 kg of Li and 1.9 kg of Co. The demand for cobalt would be 6 Mt. The demand for lithium would be 10 million tons (consistent with the value of Garcia-Olivares [GAR 12]), i.e. approximately 60% of the known reserves reported by the USGS in 2015. This would correspond to an annual production equal to ten times the current production for lithium for the next 35 years. These developments are consistent with those observed over the last three years: in

June 2014, Tesla launched its Gigafactory project, whose
maximum production capacity should be reached by 2020
with an annual production of Li-ion batteries greater than
the world production in 2013 (35 GWh/yr). Tesla plans to
increase the production of its electric cars ten-fold in 5 years
to reach 500,000 vehicles in 2020. The company also expects
to popularize the use of its *powerwall* batteries for domestic
use and the *powerpack* for business use. Other projects of the
same scale are planned in China, with stakeholders such as
CATL, which aims to see increases from 7.6 GWh to 50 GWh
in three years, or in Sweden, which is preparing the largest
battery plant (32 GWh) in Europe[9]. China has also set itself
the target of putting 5 million hybrid and electric vehicles in
circulation in five years, and India will have 800,000 vehicles
in 2016 and 6 million in 2020. The lithium requirements are
therefore likely to explode and even if recycling Li-ion
batteries is technologically feasible at a reasonable cost, the
demand for primary raw materials will continue to grow at a
very rapid rate. Other technologies, such as Na-ion batteries,
which are less powerful but allow faster charging at a lower
cost, could also be developed. As for cobalt, it is generally a
by-product of the production of other metals such as copper
or nickel (section 4.2). An increase in its production will be
possible only if the production of carrier metals increases too.
It should also be noted that cobalt is mainly refined in
China, which exposes it to the same uncertainties as rare-
earth elements.

Monitoring the electricity demand could partially cover
the need for flexibility induced by the development of
renewable energies. However, in the absence of massive
interconnections, the storage of electricity will remain
indispensable to match the demand and the production of
energy. The technologies envisaged are highly variable, such
as gravity storage, compressed air, Power to Gas and Gas to

9 http://www.usinenouvelle.com/article/les-gigafactories-creent-l-emoi-chez-leurs-fo
urnisseurs-de-matieres-premieres.N51354.

Power techniques that use electrolysis of water to produce hydrogen, possibly followed by methanation, or electrochemical storage in batteries. None of these techniques has been implemented at reasonable costs at present. It would take 12 million tons of batteries using 360,000 tons of lithium to store two days of French electricity consumption with high-performance Li-ion technology like that used in Tesla cars, while 40,000 tons of this metal are extracted every year. In France alone, the *Agence de l'Environnement et de la Maitrise de l'Energie* proposes an inter-seasonal electrical storage of 17 Gigawatts of power (equivalent to one quarter of nuclear capacity) by using "Power to Gas" technology. All the technologies envisaged will require conventional building materials such as steel (compressed air and gravity storage) or rarer elements in the case of giant batteries using vanadium or lithium-ion cells. For example, compressed air technology covering a power range of around 10 MW and a discharge time of around ten hours, which is compatible with wind generation, uses around 320 t/MW of steel, 2.5 t/MW of Al and 5 t/MW of Cu [KLO 08]. A 2 MW hydrogen storage system could contain 120 tons of steel, 2.25 tons of Cu and 0.6 tons of Al. In both cases, the electricity storage infrastructure has a base metal content of the same order of magnitude as that of the production systems reported in Figure 5.2.

4.1.6. *Energy efficiency*

Some elements are used to produce "superalloys" to improve energy performance by reducing vehicle weight and mechanical strength. Aluminum is the third most used element in the manufacture of cars, which can hold up to 10% weight, or 150 kg. To reduce the weight of vehicles, magnesium is usually alloyed with aluminum. Other more critical elements involved in the transport sector include cobalt, which is used in some high-temperature superalloys. In 2010, 16,600 tons of cobalt were used for aircraft engine alloys. In the same year, 69 tons of indium were incorporated into high strength alloys in the aircraft and automotive sectors [TAL 11]. Niobium and tantalum also make it possible to improve the high-temperature mechanical resistance of steels, which may contain between 40 and 70% niobium. Low-grade niobium steels are also used for automotive bodies, off-shore oil rigs, pipelines, and for some petrochemical applications. In 2008, 63,000 tons of niobium were used for the manufacture of steel and alloys, and in 2010, 200 tons of tellurium were used to produce high-performance steel, copper and lead alloys [TAL 11]. Tantalum, rhenium, chromium, molybdenum, nickel and tungsten, zinc, titanium, scandium and rare-earth elements are also incorporated in metal alloys in varying but significant quantities. For example, 8000 tons of rare-earth oxides were used in iron and aluminum alloys in 2010, excluding battery applications [MOR 11].

Approximately 18-20% of the world's electricity production is used for lighting, a sector that has undergone a rapid transition with the replacement of traditional incandescent bulbs with fluorescent bulbs and then light-emitting diodes (LEDs). According to the International Energy Agency, switching from incandescent lamps to fluorescent lamps would reduce global demand for electricity for lighting by 18%; but the replacement of a simple technology with a much more complex technology also involves the replacement of tungsten by gallium, indium, phosphorus, aluminum and rare-earth elements present in fluorescent bulbs and LEDs. Different dopant combinations have been developed to improve the conversion of electricity into different clear emission colors. Cerium, europium, terbium and yttrium are all used in modern LEDs.

4.2. The special cases of "high-tech" metals and co-products

It is important to note that many metals used in high-tech industries and especially in the energy sector exist only as by-products of the exploitation of primary metals which have sufficiently high concentrations to form economically exploitable deposits. Increasing the production of a by-product metal implies increasing the mining activity, which is not possible if the demand for the main metal remains constant or drops. Therefore, the increase in demand for a by-produced element may remain unsatisfied, even if there are primary reserves. Most semiconductor metals exist only as by-products, the recovery of which occurs only during processing of the carrier metal according to the metallurgist's specific industrial strategy. Bauxite (aluminum ore) is the main industrial source of gallium, zinc deposits are the main source of indium and germanium, and copper deposits are the main source of molybdenum, rhenium, selenium, and tellurium. The following are now exclusively by-products: As, Bi, Cd, Co (there is only one mine in Morocco, producing cobalt as the main ore), Ga, Ge, Hf, In, Ir, Os, Re, Rh, Ru, Sc, Se, Te, Tl, V.

A lot of research is being done to substitute rare-earth elements or reduce their consumption in permanent magnets. It is already possible to manufacture synchronous motors with permanent magnets of high efficiency without using rare-earth elements. For electric motors, one option could be the reluctance motor, which uses electromagnets instead of permanent magnets. For hybrid cars, the general trend is to recover the rare-earth elements and recycle them. The market for magnets is currently estimated to have the following distribution: 34% hard ferrites, 65% rare-earth magnets (samarium-cobalt and more recently neodymium-iron-boron (NIB)) used for electric vehicle engines and wind turbines and 1% aluminum-nickel-cobalt (Alnico) [GUT 11]. On average the NIB magnets contain 69% iron, 1% boron and 30% rare-earth elements, 70% of which comprises neodymium, 24% praseodymium, 5% dysprosium and 1% terbium (Morgan, 2011). Hybrid and electric vehicles contain about 1 kg of neodymium distributed in various permanent magnets. According to Molycorp15, a 55 kW Prius motor contains permanent magnets with Nd, Pr, Dy and Tb in the engine, more than 25 micro-electric motors powered with NdFeB magnets in adjustable seats, windscreen wipers, electric or power steering systems. In 2010, 24,060 tons of rare-earth elements were used for the manufacture of permanent magnets, including 16,700 tons of neodymium, 5,630 tons of praseodymium, 1,200 tons of dysprosium, 480 tons of gadolinium and 50 tons of terbium for NIB magnets with the following distribution: 61% in electrical and electronic equipment such as hard disks, loudspeakers and microphones, 15% in wind turbines, 14% in electric vehicles and 10% for magnetic resonance scanners. The quantities used for wind turbines and electric vehicles are increasing, while the replacement of hard drives with flash memories could loosen the constraint by reducing the demand for rare-earth elements for this use.

5

Average Material Intensity for Various Modes of Electricity Production

The previous chapter showed that major energy sectors have significant and varied mineral resource requirements. In this chapter, we attempt to quantify the need for some "structural" raw materials for electricity generation from fossil or renewable sources (concrete, steel, aluminum and copper). Our aim is to compare the requirements for these materials, which are not dependent on technological innovation but which carry the largest share of energy needed to produce the raw materials. Indeed, the variability of technologies and their rapid evolution make it difficult to quantify the full requirements for energy raw materials. Moreover, there are relatively few validated studies that exhaustively report on material intensities (in tons per MW of power or MWh of energy produced). In addition, many published studies are limited to the isolated production infrastructure of other equipment (transformers, connections, foundations, etc.).

Most wind turbine life cycle analyses, for example, report copper intensities in the range of 2 to 4 t/MW of power. However, Falconer [FAL 09] showed that the intensity rises to 8–12 t/MW for offshore wind turbine farms if the connection network between the wind turbines to the mainland is included.

Moreover, wind turbine technologies can be quite different: double-fed asynchronous generators (DFAG), conventional asynchronous generators (CAG), conventional synchronous generators (CSG) and permanent magnet synchronous generators (PMSG). These technologies require quantities of copper that range from 0.3 to 4 t/MW. Quantities of steel and concrete also differ for onshore and offshore wind turbines, depending on the type of foundation.

The same observations can be made for photovoltaic technologies on buildings or on farms, with or without tracking systems. The variability is not limited to the most common elements, and rare elements also show strong variations (rare earths in the permanent magnets of wind turbines or gallium, indium, selenium in thin-film solar panel technologies). A major difficulty in the global analysis of raw material requirements for energy comes from the diversity of the industrial sectors involved. The chain from fuel production to power generation, transmission, distribution, storage and use is long and complex (Figure 5.1).

An overall evaluation must dive into the details of this chain and accurately identify the requirements for each sector. In the following, we focus on electricity generation only and we emphasize the partial nature of identified needs, which cover only a portion of the overall needs.

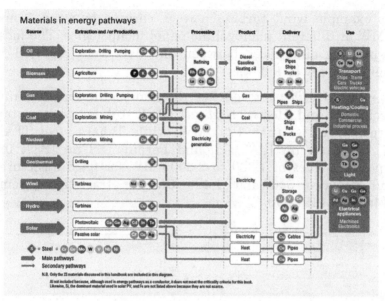

Figure 5.1. *Main requirements of raw materials for energy production [ZEP 14]. For a color version of this figure, see www.iste.co.uk/vidal/energy.zip*

Several studies have shown that the material intensity for electricity generation infrastructure is more significant for wind and solar energy than for fossil fuel power plants ([VID 13, KLE 11, KLE 12, HER 15]). We describe an initial comparison for three data sets in Figure 5.2 and Table 5.1 for the concrete, steel, aluminum and copper in 11 different power generation technologies. The material intensity values for renewable energy technologies show much greater dispersion than fossil fuel power plants. Part of this dispersion can be explained by the fact that we considered studies published between 1995 and 2014 and that technologies evolve very rapidly. Another reason stems from the measurement itself, which in some cases considers the solar panel or isolated wind turbine while other studies integrate the interconnections and/or the distribution of electricity to the grid. A final reason is the different technologies within each type identified in Figure 5.2, with for example wind turbine powers varying between 1 and 6 MW or different photovoltaic panel technologies. The most significant differences between the data sets reported in Figure 5.2 are as follows:

– *Concrete in photovoltaics:* the material intensities from Ecoinvent and Hertwich *et al.* [HER 15] for photovoltaic farms were less than 100 t/MW, while Tahara *et al.* [TAH 97], Pacca *et al.* [PAC 02] and Perpinan *et al.* [PER 09] reported values of 1,300 to 2,300 t/MW for fixed structures and between 540 and 1,400 t/M for tracking systems ([MAS 06, PER 09]).

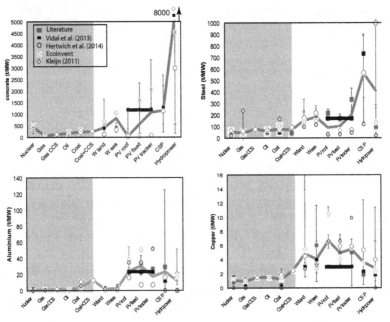

Figure 5.2. *Quantities of four structural raw materials used to manufacture different power generation infrastructures. Error bars indicate the range of values covered by the compiled data. The data labeled "Hertwich et al. (2015)" are an average of the values reported by these authors for different regions of the world and different technologies. The blue line is the average of data from Hertwich et al. [HER 15], Ecoinvent and the literature, as used by Vidal et al. [VID 13] (black squares) that were completed (red squares). Fossil fuel power generation technologies are in the gray shaded area. For a color version of this figure, see www.iste.co.uk/vidal/energy.zip*

– *Concrete in dams:* the average concrete intensity calculated from all country values used by Hertwich *et al.* [HER 15] was about 3,000 t/MW, while that obtained from Ecoinvent data was 8500 t/MW, that of Pacca *et al.* [PAC 02] was 7,600 t/MW and that from Rule *et al.* [RUL 09] was 6,700 t/MW.

– *Iron in hydroelectric structures:* the average iron intensity calculated from the values used by Hertwich *et al.* [HER 15] was very high (1,000 t/MW) and represented a third of the concrete intensity, which is not representative of dams but rather of "run-of-river" technologies. In comparison, Ecoinvent data was 135 t/MW, which is reasonably in agreement with the average of values calculated in this study (96 t/MW).

– *Iron in photovoltaics:* data from Kleijn *et al.* [KLE 11], Hertwich *et al.* [HER 15] and Ecoinvent were below 70 t/MW, while Azzopardi and Mutale [AZZ 10] and Raugei *et al.* [RAU 07] reported values around 200 t/MW for roof photovoltaics, and Perpinan *et al.* [PER 09] reported between 200 and 430 t/MW for photovoltaic farms.

– *The aluminum intensity of photovoltaics in farms* was less than 16 t/MW according to Hertwich *et al.* [HER 15], whereas the averages calculated from Ecoinvent data were greater than 40 t/MW and those calculated from the literature were 34 t/MW. The values reported by Kleijn *et al.* [KLE 11] were 35 to 52 t/MW.

The above comparisons show that contrasted material intensities are reported in different studies, particularly for renewable energy technologies. The blue curve in Figure 5.2 shows the "average of averages" obtained with the values from Hertwich *et al.* [HER 15], Ecoinvent and the "literature data". This average of averages confirms that, for the same nominal installed capacity, the quantities of raw materials required to build solar and wind power infrastructures of

electricity production are generally higher than those used
by fossil fuel power plants.

Comparing material intensities per unit of produced
energy is even more telling, since it takes the production
efficiency of each generating source into account, as well as
their availability over time (Figure 5.3). This comparison
shows, for example, that for the same installed capacity, the
steel intensity in t/MW is 2 to 3 times higher for onshore
wind turbines than for coal-fired power plants. It is 5 times
higher in t/MWh, considering a yield of 0.24 and a lifetime of
20 years for wind power, and a yield of 0.33 and a lifetime of
30 years for coal (values used by [HER 15]).

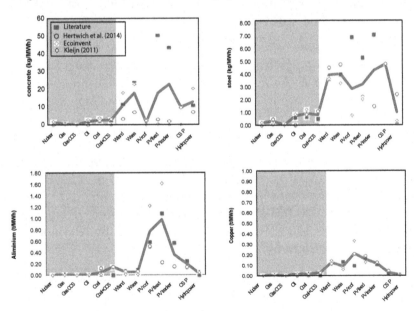

Figure 5.3. *Mass of material in kg required to produce 1 MWh electricity,*
calculated with the material intensities shown in Figure 5.2 and Table 5.1. The
gray shaded area indicates fossil fuel-based electricity production. For a color
version of this figure, see www.iste.co.uk/vidal/energy.zip

	Nuclear	Gas	Gas + CCS	Oil	Coal	Coal + CCS	Wind onshore	Wind offshore	PV roof	PV farm, fixed	PV farm, tracking	CSP	Hydro	Run-of-river	Geothermy	Biomass
Yield	0.60	0.50	0.50	0.33	0.33	0.33	0.22	0.26	0.12	0.12	0.18	0.45	0.60	0.60	0.15	–
lifetime	60	30	30	30	30	30	20	20	30	30	30	30	80	40	40	–
kgCO2/KWh (see chapter 6)	0.020	0.500	0.200	0.821	0.900	0.200	0	0	0	0	0	0	0	0	0	–
Concrete (t/MW)																
Vidal et al. (2013)	532	81	–	–	180	–	400	400	1186	1186	1186	1289	7644	672	1892	–
Average of literature data	307	72	–	–	157	198	440	1083	–	1590	2057	1171	4540	672	1892	159
min	176	38	–	–	74	198	298	1083	–	63	1281	218	552	218	1892	159
max	485	98	–	–	195	198	565	1083	–	2350	3342	2242	8823	672	1892	159
Hertwich et al. (2014)	–	91	130	71	205	269	129	322	70	92	92	1136	2989	–	–	–
max	–	132	0	71	314	–	165	620	73	93	93	2694	6043	–	–	–
min	–	51	0	71	137	–	102	198	57	87	87	562	162	–	–	–
Ecoinvent	546	41	–	242	325	–	696	1038	–	10	–	–	8498	25893	611	–
min	444	36	–	242	250	–	417	1038	10	10	–	–	7794	15626	154	–
max	676	43	–	242	400	–	1640	1038	10	–	–	–	9201	36160	1067	–
Average of averages	427	68	130	157	229	233	422	814	35	564	1075	1154	5342	13283	1251	–
MI Chapter 6 (t/MW)	427	68	130	193	193	233	618	618	427	427	427	1154	7500	0	0	–
MI Chapter 6 (kg/MWh)	1.35	0.52	0.99	2.22	2.22	2.69	16.03	13.57	13.55	13.55	9.03	9.76	17.84	0.00	0.00	–
Steel & iron (t/MW)																
Vidal et al. (2013)	49	59	–	–	52	–	130	130	169	169	169	731	0	–	–	–
Average of literature data	44	56	–	–	56	42	138	181	217	167	335	560	96	213	584	0
min	30	3	–	–	40	42	118	110	200	128	229	138	25	213	584	0
max	68	250	–	–	68	42	222	251	250	205	431	890	284	213	584	0
Hertwich et al. (2014)	–	61	77	51	72	93	173	215	24	69	69	565	2811	–	–	–
max	–	78	0	51	103	–	334	277	35	80	80	998	–	–	–	–
min	–	44	0	51	48	–	107	171	19	60	60	350	259	–	–	–
Ecoinvent	67	13	–	73	106	–	139	148	–	64	–	–	135	862	129	–
min	61	6	–	73	92	–	125	0	18	64	–	–	124	274	67	–
max	73	27	–	73	120	–	234	0	35	–	–	–	146	1450	190	–
Average of averages	55	43	77	62	78	93	150	181	88	100	202	563	410	538	356	0
MI Chapter 6 (t/MW)	55	43	77	70	70	93	166	166	120	120	120	563	410	538	356	0
MI Chapter 6 (kg/MWh)	0.17	0.33	0.59	0.81	0.81	1.08	4.30	3.64	3.79	3.79	2.53	4.76	0.97	2.56	6.78	–

Table 5.1. *Material intensities used for Figures 5.2 and 5.3. The MI Chapter 6 values are the material intensities used for dynamic modeling presented in Chapter 6. The mass of CO_2 emitted per kWh of produced electricity is from Chapter 6*

Aluminum (t/MW)

Vidal et al. (2013)	1.5	0.3	-	-	0.6	-	3.0	3.0	23.0	23.0	23.0	11.9	0.1	-	-	-
Average of literature data	0.1	0.3	-	0.2	0.5	-	2.8	4.2	18.1	34.1	26.7	29.0	0.2	0.0	21.8	1.3
max	0.2	1.1	-	1.0	0.3	-	5.3	7.5	23.0	43.2	28.0	125.0	0.9	0.0	21.8	1.3
min	0.0	0.0	-	0.2	0.3	-	0.9	1.0	15.2	19.0	23.0	0.0	0.0	0.0	21.8	1.3
Hertwich et al. (2014)	-	1.3	1.7	-	10.5	12.4	2.6	3.0	15.9	7.1	7.1	16.8	19.9	-	-	-
max	-	1.7	0.0	-	13.5	-	4.9	5.8	17.4	8.0	8.0	25.5	51.0	-	-	-
min	-	0.9	0.0	-	8.1	-	1.6	1.9	14.4	6.4	6.4	11.3	6.9	-	-	-
Ecoinvent	0.1	0.9	-	0.6	0.8	-	0.5	0.4	38.3	50.9	-	-	16.5	-	-	-
min	0.0	0.8	-	0.6	0.6	-	0.1	0.0	28.1	50.9	-	-	0.0	-	0.5	0.0
max	0.2	1.1	-	0.6	1.0	-	1.0	0.0	54.5	50.9	-	-	0.0	-	0.5	0.0
Average of averages	0.1	0.8	1.7	0.6	3.9	12.4	2.0	2.6	24.1	30.7	16.9	22.9	12.2	0.0	11.2	0.0
MI Chapter 6 (t/MW)	0.1	0.8	1.7	0.6	2.3	12.4	2	2	24	24	24	23	12	0	11	0
MI Chapter 6 (kg/MWh)	1.35	0.52	0.99	2.22	2.22	2.69	16.03	13.57	13.55	13.55	9.03	9.76	17.84	0.00	0.00	-

Cuivre (t/MW)

Vidal et al. (2013)	1.5	0.3	-	-	0.3	-	3.0	4.0	3.0	3.0	3.0	2.3	1.9	-	-	-
Average of literature data	0.8	0.8	-	0.0	1.1	-	4.6	6.0	3.0	5.0	5.0	1.9	2.0	3.5	4.0	0.0
max	1.5	1.1	-	0.0	0.5	-	20.1	11.7	3.0	7.5	5.1	3.2	6.6	3.5	4.0	0.0
min	0.2	0.5	-	0.0	-	-	1.6	0.9	3.0	3.0	0.3	0.4	0.1	3.5	4.0	0.0
Hertwich et al. (2014)	-	1.1	1.5	-	1.9	2.5	4.5	3.4	6.5	5.9	5.9	5.4	11.4	-	-	-
max	-	1.5	0.0	-	2.7	-	5.7	6.6	7.5	6.9	6.9	12.4	11.4	-	-	-
min	-	0.8	0.0	-	1.4	-	3.5	2.1	5.6	4.8	4.8	2.5	1.0	-	-	-
Ecoinvent	1.3	0.9	-	1.5	1.4	-	5.5	2.7	10.5	4.0	-	-	2.3	-	-	-
min	0.9	0.8	-	1.5	1.3	-	1.5	0.0	10.1	4.0	-	-	2.3	-	4.0	0.0
max	1.5	1.1	-	1.5	1.5	-	13.9	0.0	11.4	4.0	-	-	2.3	-	4.0	0.0
Average of averages	1.1	0.9	1.5	1.4	1.3	2.5	4.9	4.1	6.7	5.0	5.4	3.6	2.8	3.5	3.8	0.0
MI Chapter 6 (t/MW)	1.1	0.9	1.5	1.4	1.3	2.5	7.0	7.0	4.5	4.5	4.5	4.0	0.1	3.8	4.0	0.0
MI Chapter 6 (kg/MWh)	1.35	0.52	0.99	2.22	2.22	2.69	16.03	13.57	13.55	13.55	9.03	9.76	17.84	0.00	0.00	-

Table 5.1. *Continued*

5.1. Innovation leads to new needs

In their report "Beyond the supercycle: how technology is reshaping resources", Mckinsey's Global [1] states that fundamental technological changes are taking place, which are reshaping both consumption and resource production. Technological advances in data analysis, artificial intelligence and robotics are announced as a revolution that will lead to substantial gains in energy efficiency. The authors' optimism is reassuring, but their conclusions must be qualified. Indeed, the evolution of performance by technological improvement is associated with a decrease in costs, prices, and ultimately with an increase in users and therefore consumption (bounce-back effect). In addition, technological improvement is associated with an increasing complexity of products and materials and does not systematically reduce material intensity. For example, the latest generation of wind turbines with a capacity of 6 MW are 170 m high and contain about 1,500 tons of steel when anchored to the ground (mast + nacelle + foundations, footbridges, etc.). These quantities correspond to 250 tons of steel/MW power, which is in the high range of values shown in Figures 4.4 and 4.5 and Table 4.1. New technologies are being considered to mount offshore wind turbines on Windfloat[2]-type floating platforms (Figure 5.4), which allow them to be installed in deeper areas and therefore farther from the coast. These platforms contain 1,200 to 1,800 tons of steel, bringing the steel intensity of the production infrastructure to 450 tons/MW, which is more than twofold the value used to construct Figures 4.4 and 4.5. In addition, there are requirements for electricity storage and distribution (about 200 t/MW of steel for compressed air or

---

1 http://www.mckinsey.com/business-functions/sustainability-and-resource-productivity/
our-insights/how-technology-is-reshaping-supply-and-demand-for-natural-resources.
2 http://www.windpowerengineering.com/construction/floating-turbine-platform-ready-to
-tap-the-2-tw-offshore-potential. http://www.antoniovidigal.com/drupal/cd/WindFloat-Qu
ase-há-um-ano-no-Mar.

gas to power systems). Many other raw materials that are considered abundant and cheap today could become critical tomorrow if they are used *en masse* in large-scale implemented technologies. This was discussed in the previous chapter for lithium, for which the demand is expected to explode in the near future, but other elements that are not expected to experience a supply risk may also be affected. For example, magnesium is a promising component in the storage of H_2 as magnesium hydrides. In France, the McPhy Energy company is a leader in the development of technological solutions for storing hydrogen in the form of MgH_2. Although it requires significant technological development before it can be operational, one application is the mobile storage of hydrogen to power a vehicle equipped with a fuel cell. Such a cell with an autonomy of 500 km would contain 70 kg of Mg. It would take about 4 million tons of Mg to equip 10% of the existing vehicle fleet. Other uses of Mg are also planned in the automotive sector to reduce the weight of vehicles and thus reduce their fuel consumption. In 2009, fuel consumption was 5 to 20 kg per vehicle, but a Volvo concept car uses 50 kg of magnesium in the wheels, chassis and engines. Currently, 70% of Mg is produced in China through the decarbonation of dolomite. The process consumes 10,000 kWh/t and produces 42 kg eq-CO_2/kg Mg. Other production methods are possible, such as the electrolysis of molten $MgCl_2$ salt extracted from seawater or brine (18,500 kWh/t), but the current total production capacity remains around 4 million tons per year, of which 1 million tons per year are for current uses. According to Clark & Marron, demand for magnesium is expected to grow by an average of 7.9 % per year until 2019[3]. Magnesium, which the European Union[4] lists as a "critical" element, could see its criticality level rise sharply with supply risks if H_2 storage technology in Mg hydrides were to be generalized

---

3 http://alliancemagnesium.com/fr/alliance-magnesium-participera-a-la-conference-74americaine-dinvestissement-fsx-interlinked.

4 http://europa.eu/rapid/press-release_MEMO-14-377_en.htm.

to nomadic applications (which remains very unlikely for the time being). Another example in the thermal storage sector is nitrate salts. A 50 MW solar power plant currently requires 28,000 tons of nitrate salts to store heat. To meet the IEA's 2050 targets of 10% of the world's electricity being produced from concentrated solar energy[5], about ten times the current global production of nitrate salts each year would need to be used.

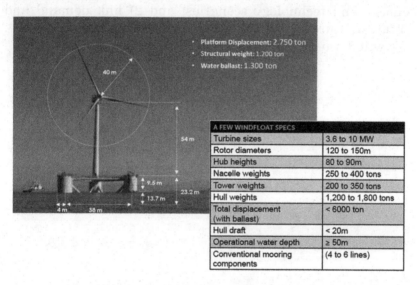

Figure 5.4. *Floating platform for offshore wind turbine and technical specifications*

Considerable work still needs to be done to build life-cycle inventories for the use and production of raw materials from a dynamic foresight perspective. The compilation of data on energy intensity and environmental impact associated with the primary and secondary production of materials required for energy transition should cover high-tech metals for which our historical knowledge of natural deposits and

5 Energies: comment les stocker? Journal du CNRS N°271 (2013), vol. 20. http://www2.cnrs.fr/sites/communique/fichier/3_jdc_271_1_enquete_sp.pdf.

concentration processes, as well as industrial separation processes, remains limited. However, the analysis should not be limited to high-tech metals, as the volumes of base metals used for energy capture, storage and distribution are enormous, and they bear the energy bill (see Chapter 3). Long-term projection of demand and production capacity requires the development of dynamic models in order to 1) compare material and energy requirements for different energy and technology scenarios; and 2) link demand and future resource availability for imposed demand scenarios. We will discuss two examples in the next chapter.

6

Dynamic Modeling

In the previous chapter, we saw that major energy production infrastructures are needed to capture and convert renewable energies (which are diluted flow energies) into electricity. On average, these infrastructures have higher material intensities than fossil fuel power plants, and their production is not neutral because they require significant amounts of energy to build (Chapter 3). In order to assess the material and energy impacts of existing energy transition scenarios, it is important to estimate the requirements first. Beyond the cumulative quantities (stock) of raw materials that are needed to carry out a given scenario, other constraints may appear, such as annual consumption (flow), which partly depends on the speed of transition. Estimating the gains and impacts of a scenario is therefore a dynamic problem. Initially, constructing a new energy infrastructure is a source of over-consumption of raw materials and fossil energy sources because currently, most energy produced comes from fossil fuels. However, over time, construction will use more and more renewable energy and recycled metals, the proportion of which increases at the same rate as primary production, with a time lag that corresponds to the life span of manufactured products. These products include those used in the field of energy, and if energy transition scenarios are built by 2050, the first wind and solar power generation infrastructures will have to be replaced within 20 to 30 years because they will have a life span of 20 to 30 years. Finally, fossil fuel power stations can be dismantled and their metals can be recycled as low-carbon electricity generation becomes more significant. All these aspects must be integrated into dynamic models that allow the coupling of fossil fuel and renewable energy, primary raw materials and recycling.

Another issue is the adequacy of future raw material supply needs and production capacities. In the early 1970s, Meadows *et al.* [MEA 72] suggested that many mineral resources would be exhausted in the short term. This has not been the case, but the idea of a widespread shortage of fossil fuel resources before the end of the century is still often expressed. It is supported by approaches such as the one developed by Hubbert [HUB 56] some 60 years ago to predict the evolution of conventional fossil fuel hydrocarbon production in the United States. The possibility of primary metal scarcity within the 21st Century is considered using the Humbert approach [LAH 10], based on dynamic models [SVE 14a], or geological arguments and the future mining production capacities [NOR 14]. A recurrent limitation of certain studies on the future of fossil fuel resources is that reserves constitute a static stock and that the time to full depletion could be estimated by applying an annual consumption rate. In reality, reserves evolve based on geological availability as well as economic and technological conditions. A century ago, copper was mined from 4% deposits, while today it is mined at a lower cost from deposits with a concentration of < 1%. Since the volume of deposits containing less than 1% copper is immensely greater than that of deposits containing 4% copper, reserves therefore increase with technological improvements at constant prices. The price depends on demand, which in turn depends on the standard of living, population, technology and economic conditions. Again, these aspects can and should be integrated into dynamic models.

Two models are presented below: the first considers the evolution of raw material and energy requirements for the three scenarios already mentioned in Chapter 5. The second model analyzes the evolution and coupling between copper production – a very significant metal in the energy transition – demand, production costs and prices, using prey–predator dynamics derived from biology.

6.1. Raw material and energy requirements for three global energy scenarios

In this chapter, we propose a dynamic model for material stocks and flows, and energy flows according to three scenarios: Bluemap, Ecofys-WWF and Garcia-Olivarès *et al.* [GAR 12] (Figure 6.1). The IEA Bluemap scenario was selected because it is somewhat conservative in anticipating that about 40% of electricity production will be from renewable sources by 2050. The WWF scenario predicts a major deployment of renewable energy sources, which would completely replace fossil fuel-based electricity generation by 2050. Garcia-Olivarès' scenario is the most demanding since it assumes that all the energy consumed in the world in 2030 will be produced by a mixture of wind turbines, concentrated solar energy (CSP) and hydropower. This unrealistic scenario has the advantage of providing maximum values for material needs. Contrary to Garcia-Olivarès *et al.* [GAR 12], we have assumed the energy mix to be completely deployed by 2050 rather than 2030. As in Chapter 5, our study focuses on the steel, concrete, aluminum and copper requirements, which are the main and non-substitutable raw materials. Their historical production has been well documented for many decades and their future needs can be estimated relatively reliably because they are not very dependent on technological innovation (unlike rare and small metals).

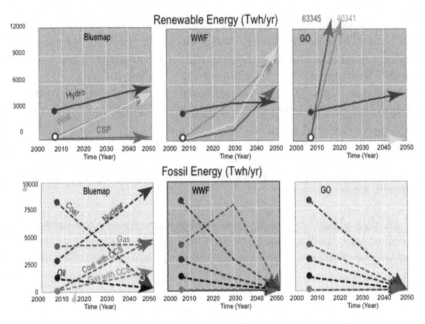

Figure 6.1. *Evolution of the energy mix for three scenarios: Bluemap (IEA 2010), Ecofys-WWF [ECO 12] and Garcia-Olivarès et al. (GO) [GAR 12]. For a color version of this figure, see www.iste.co.uk/vidal/energy.zip*

6.1.1. Description of the energy-raw materials model

6.1.1.1. Electricity generation infrastructure

The value chain from primary extraction, use, end-of-life and recycling has been modeled in a simplified manner in Figure 6.2. Three types of power generation infrastructure have been considered, which are defined by the life span of the facilities. We have made the simplistic assumption that the life span of electricity generation infrastructures from wind and solar power sources is 20 years, while coal, gas and oil-fired power plants are 40 years, and nuclear power plants are 60 years. For each type of installation, the stock in tons of *"primary metal in the infrastructure"* is supplied by the flow in tons/year of *"primary production"*, and the stock of *"recycled material in the infrastructure"* is supplied by the *"recycling"* flow. This is of course the same infrastructure of energy generation – the two stocks simply identify respective shares of primary and recycled raw material contained within it. The distinction between primary and recycled metals allows the energy required to produce the raw materials in the infrastructure over time to be quantified, taking into account the differences between primary and recycled raw materials. The part of the infrastructure manufactured at t0 reaches its end of life in t0 + Δt, where Δt = lifetime (20, 40 or 60 years depending on the fuel). The stock *"primary metal in the infrastructure"* is then reduced by the outgoing flow of *"primary metal in EOL products"*. This t0 + Δt flow is equal to the *"primary production"* flow at t0 multiplied by a rate of collection and recycling (R). The rest of the material is lost at t0 + Δt through *"loss of primary metal"* = *"primary production"* x (1 - R). The *"primary metal in EOL products"* flow feeds a second stock of *"EOL productsl"*, which is also supplied by the already-recycled metal flow entering a new recycling loop (*"recycled metal in EOL products"*). At t0 + Δt, this *"recycled metal in EOL products"* flow is equal to the *"recycling"* flow at t0 multiplied by R. The third outflow (Oil-Coal-Gas and Nuclear) or inflow (Solar and Wind) from the *"EOL products"* stock is the metal that comes from dismantling the fossil fuel power plants. Indeed, some scenarios require fossil fuel power stations to be dismantled before the end of their useful life (Ecofys-WWF and Garcia-Olivares), while others require a significant proportion of fossil fuels to be conserved in the energy mix (Bluemap). Depending on how the capacity of the fossil fuel power plant evolves, any surplus metal resulting from the dismantling is automatically recycled to build the renewable energy power plant. For the scenario being considered, fossil fuel power plants are dismantled if there is no primary production and no recycling. Otherwise, the materials are recycled for the same use when the plants reach the end of their useful life. This amounts to imposing the following conditions:

– if the oil-coal-gas *"primary production"* = 0, then *"dismantling"* > 0 and *"recycling"* = *"primary metal in EOL products" + Recycled metal in EOL products" – "dismantling"* = 0. In this case, the *"Recycled metal in EOL products"* flow is redirected to the *"EOL products"* stock of the solar and wind energy sector when the plants reach the end of their useful life. The raw materials are recycled instantaneously, so that the *"EOL products"* stocks are empty;

– otherwise if oil-coal-gas *"primary production"* > 0 (maintenance or growth of end-of-life power plants), then *"dismantling"* = 0 and *"recycling"* = "primary metal in EOL products" + "dismantling" + "Recycled metal in EOL products" > 0. In this case, primary production is positive if recycling is not sufficient to maintain the installed capacity.

The model presented in Figure 6.2 is obviously an extreme simplification of reality. In particular, it assumes that the metals in the infrastructure are recycled for the same use (they go back to the power generation infrastructure). In reality, end-of-life products are not recycled according to their use. However, our assumption makes it possible to take into account the infrastructure renewal in a simple way, both during its deployment (before 2050) and during the potential dismantling of fossil fuel power plants. Above all, this assumption makes it possible to account for the different energy requirements of the primary production and recycling over time. The model begins in 1900 and ends in 2100, with the following assumptions:

– installed capacities of nuclear power stations increase linearly from 0 in 1950 to 2010;

– installed capacities of other fossil fuel power stations and hydropower plants increase linearly between 1900 and 2010;

– installed capacities remain constant after 2050.

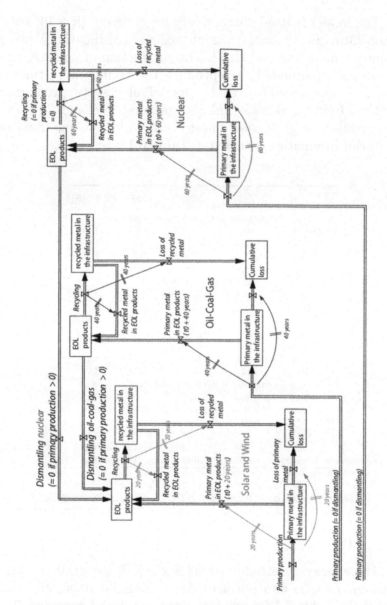

Figure 6.2. *For a color version of this figure, see*
www.iste.co.uk/vidal/energy.zip

The model is used successively for concrete (R = 0), steel, aluminum and copper in each scenario using the Vensim dynamic modeling software[1]. The production of primary raw materials is adjusted to reproduce the change over time of energies generated by each type of electricity production method (wind and solar, oil-coal-gas and nuclear). The raw materials and generated energy are linked using the raw material intensities reported in Table 6.1.

	2010	2030	2050	2050 / 2010
Primary energy of production (MWh/t)				
Concrete	0.3	0.2	0.2	0.84
Steel	6.11	5.11[1]	4.21[1]	0.68
Al	23.6	20.8	18.1	0.77
Cu	8.3	8.3	8.3	1.00
Recycling energy (MWh/t)				
Steel	1.9	1.5	1.1	0.57
Al	2.4	2.1	1.8	0.77
Cu	4.2	4.2	4.2	1.00
emitted CO_2 (t/t of primary production)				
Concrete	0.3	0.225	0.15	0.50
Steel	2.3	1.535	0.77	0.33
Al	12.2	9.1	6	0.49
emitted CO_2 (t/t of recycled metal)				
Steel	0.6	0.4	0.2	0.33
Al	1.4	1.05	0.7	0.50
Global recycling rate R (%)				
Steel	0.65	0.67	0.68	0.73
Al	0.48	0.51	0.54	0.60
Cu	0.43	0.47	0.51	0.60

Table 6.1. *Energy consumed and CO_2 emitted in the production of concrete, steel, aluminum and copper, and the recycling rate. Production energies are derived from Birat et al. [BIR 14]*

The energy required to produce the raw materials and the amount of CO_2 emitted are also calculated using the unit values in Table 6.1 for primary production and recycling. The energy used to produce concrete, steel, aluminum and copper

---

1 https://vensim.com.

is expressed in electric TWh of consumed energy (secondary) in Table 6.1. The values take the foreseeable improvement in energy efficiency into account (Chapter 3). Finally, the collection and recycling rates (R) for raw materials in the electricity generation infrastructure have all been set at 70% except for concrete, which is not recycled (or at least, not for the same use). The model is limited to electricity generation only, but additional requirements for storage, distribution and use of electricity are briefly discussed below.

6.1.1.2. *Global production*

In a second step, the requirements for energy infrastructure are compared to worldwide production using a model similar to that in Figure 6.2, but which considers all consumer goods that contain the raw material of interest (Figure 6.3). An average 30-year life span is considered for global iron (steel), copper and aluminum, and the aggregated collection x recycling rates (R) are those estimated for all goods at the world scale. The value of R for *steel* was estimated to replicate the historical USGS-reported primary iron (cast iron) data and recycled scrap metal data reported by the AME[2] group (Figure 6.6). Future steel production up to 2050 corresponds to the AME baseline scenario, which is similar to the Bluemap scenario estimates. Primary production is expected to increase from 1,300 Mt in 2007 to 2,200 Mt in 2050 and recycling from 400 Mt to 1,000 Mt. To replicate these developments, the R for primary steel has to increase from 63% in 2000 to 73% in 2100. Future primary *copper* production has been estimated by Northey *et al.* [NOR 14]. The value of R for copper was set at 40% for the period 1900 to 2015 to replicate the annual amount of copper produced from end-of-life products since 1960 and primary production since 1900, as reported by the USGS (Figure 6.6). Reliable data for *aluminum* are more difficult to come by and its recycling rate is considered to be equal to that of copper. A

2 http://www.ame.com.au/Website/FeatureArticleDetail.aspx?faId=12.

constant growth rate of production was assumed to reach 150 Mt/yr in 2050, in line with [IEA 10a] estimates that range betwwen 142 and 190 Mt/year. For *cement*, the annual production was assumed to increase to 7 Gt/year in 2050, which is slightly higher than the IEA estimates (5 to 6 Gt/year).

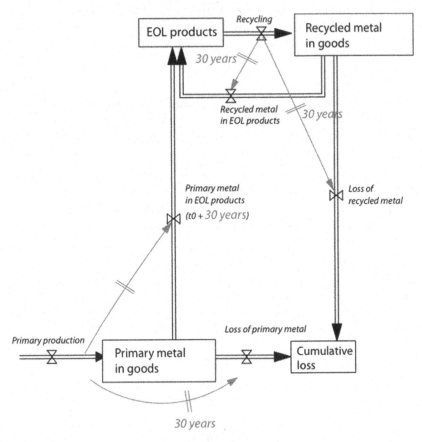

Figure 6.3. *Diagram of the model used for the dynamic modeling of raw material flows and stock on a global level. Recycling rates are shown in Table 6.1 and an average life span of 30 years is used for all materials*

6.1.2. *Calculated raw material and energy requirements for future electricity-generation infrastructure*

6.1.2.1. *Copper consumption*

For example, the annual flows and cumulative amount of copper needed to build the power-generation infrastructure in the three scenarios are illustrated in Figure 6.4.

For the Bluemap scenario, the calculated stock of copper in the electricity-generation infrastructure is about 40 Mt in 2050 (blue box, black line in the graph in row *"stock in infrastructure"* and column *"total"* (Figure 6.4), which is higher than the 29 Mt estimated by Hertwich *et al.* [HER 15] for the same scenario. Moreover, for a 20-year life span of wind turbines, PV and CSP, the infrastructure installed between 2000 and 2050 must be replaced before 2050. With each recycling loop, 30% of the copper is lost (R = 70%), so that the actual cumulative amount of copper to be produced exceeds 50 Mt in 2050 and continues to increase after this period to maintain the infrastructure. In 2050, 60% of the copper incorporated in the energy production infrastructure is of primary origin, compared to 90% in 2020. Figure 6.4 shows that annual primary consumption peaks at 1.2 Mt/year at the end of the deployment period and drops to 0.5 Mt/year for infrastructure renewal after 2050. The needs are mainly due to the development of renewable energy production facilities, which use up 32 Mt of the 40 Mt total (black line in the graph in row *"stock in infrastructure"* and column *"wind and solar"* in Figure 6.4). Copper consumption for oil-coal-gas power plants is twice as low as that of wind and solar power generation, and the contribution of nuclear power plants is negligible.

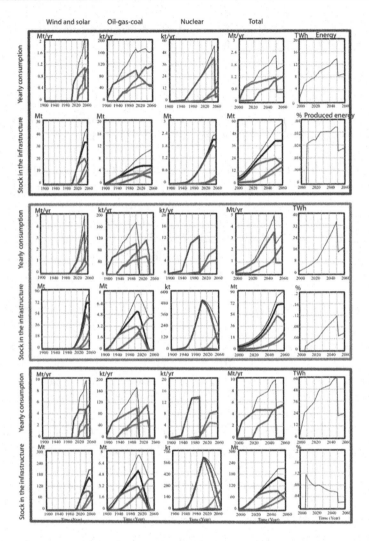

Figure 6.4. *Annual copper consumption and cumulative stock in electricity generation infrastructure for the Bluemap scenario (blue box), WWF-Ecofys (green box) and Garcia-Olivarès et al. (red box). The red lines show primary copper, the green lines show recycled copper, the gray lines show lost copper, the blue lines show primary copper + recycled copper + lost copper, the black lines show the cumulative amount of primary and secondary copper in the infrastructure. The two graphs on the right show the annual amount of energy for primary production and recycling, and the corresponding percentage of energy produced by the infrastructure, respectively. For a color version of this figure, see www.iste.co.uk/vidal/energy.zip*

For the WWF-Ecofys scenario, if we include losses associated with recycling, the cumulative amount of copper in 2050 is about 80 Mt, of which 70 Mt is in the infrastructure (red box, black line in the graph in row *"stock in infrastructure"* and column *"total"* in Figure 6.4). Ecofys-WWF [ECO 14] estimated that their scenario requires about 2 Mt/year of copper between 2010 and 2050. In fact, the sum of primary + recycling + lost + primary fluxes reaches 5 Mt/year in 2050 and primary copper consumption increases from 1 to 3.3 Mt/year between 2025 and 2050 (according to our very optimistic hypothesis that 70 % of primary copper is recycled). This amount of primary copper is equivalent to nearly a quarter of the world's 2010 production, or the cumulative 2012 production of the world's eight largest copper mines: Andina, Escondida, El Tiente, Los Bronces, Radomiro Tomic and Collahuasi (Chile), Toquepala and Cerro Verde (Peru) and Cananea (Mexico). The increase in copper consumption for electricity-generation infrastructures alone (from 1 to 3.3 Mt/year between 2025 and 2050) is identical to that observed worldwide between 1970 and 2000 for all uses of copper. As the stock of copper in fossil fuel power plants in 2010 (5.5 Mt) is well below the cumulative stock of solar and wind power plants in 2050 (70 Mt), the material flows from the dismantling of fossil fuel power plants remain well below the annual requirements for renewable energies.

For the Garcia-Olivares et al. scenario [GAR 12], the cumulative amount of copper reaches 160 Mt in 2050, in fair agreement with our own estimates (183 Mt). Approximately 50 Mt of copper is lost between 2010 and 2050 and the total amount to be produced is 210 Mt. The quantities of raw and recycled copper up to 2050 represent respectively 6 and 22 times the world production in 2010. Annual primary copper consumption for energy generation infrastructure is about 4.5 Mt/year between 2020 and 2050, which corresponds to over a quarter of the global production in 2010.

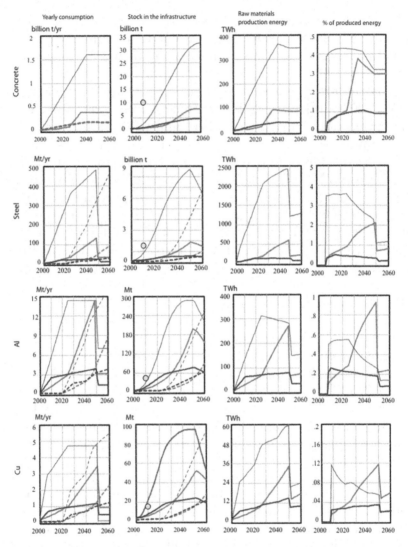

Figure 6.5. *Annual consumption (first column) and cumulative requirements (second column) of raw materials (solid lines) and recycled materials (dotted lines) for the deployment of the energy production infrastructure. The next two columns show the annual energy required for primary production and recycling, and the proportion of this energy produced by the energy-generation infrastructure. Red line: Garcia-Olivarès et al. [GAR 12], green line: Ecofys-WWF [ECO 12] and blue line: Bluemap [IEA 10b]. Yellow dots: annual global production. For a color version of this figure, see www.iste.co.uk/vidal/energy.zip*

The annual consumption of recycled copper between 2040 and 2070 for the infrastructure exceeds the global amount of recycled copper from all end-of-life products in 2010. The quantities of raw materials reported in Figure 6.4 are for energy generation only. Garcia-Olivarès *et al.* [GAR 12] estimated that the share of copper in the energy production infrastructure only represents 55% of the total amount required to shift from the current energy mix to a mix based on renewable sources. The cumulative raw copper production up to 2050 could therefore reach 300 Mt for this scenario (330 Mt estimated by Garcia-Olivarès *et al.* [GAR 12]) and 100 Mt for the Ecofys-WWF scenario. This would represent 6 to 20 years of global copper production in 2010, or between a sixth and half of the world's known copper reserves in 2010.

6.1.2.2. *Concrete, steel and aluminum consumption*

Cumulative production and consumptions of concrete, steel and aluminum show similar trends to copper (Figure 6.5). The annual consumption of primary aluminum between 2037 and 2050 is equivalent to 30 to 40% of the world production in 2010 for the Ecofys-WWF scenario and 22% between 2025 and 2050 for the Garcia-Olivarès scenario. The annual consumption of primary steel between 2030 and 2050 accounts for about a third of global production in 2010. In comparison, the Bluemap scenario consumptions are much lower. For all three metals, the amounts of metals recycled from the infrastructure increase rapidly over time. From 2050 onwards, the annual recycling of the energy production infrastructure according to the Ecofys-WWF and Garcia-Olivarès scenarios is equivalent to the 2010 global recycling of copper and aluminum from all end-of-life products. For the Bluemap scenario, the infrastructure annual recycling is equivalent to half the global recycling in 2010. These results assume that widespread growth in the current recycling infrastructure will deal with the growing flows of metals to be recycled.

6.1.2.3. *Energy and emitted CO_2*

The annual energy required to produce primary and recycled raw materials for the electricity-generation infrastructure is shown per material in Figures 6.4 and 6.5. For all three scenarios, the annual energy required to produce concrete, steel, copper and aluminum increases until 2050 and then decreases sharply thereafter (Figure 6.7), as we have assumed that the installed capacity remains constant after 2050. For the Garcia-Olivarès scenario, this energy rises from 63 TWh/year in 2000 to 3,050 TWh/year in 2050, before stabilising at 1,650 TWh/year after 2050. The energy used to produce the raw materials is thus multiplied by 50 between the years 2000 and 2050. It is 25 times higher after 2050 than in 2000, even if the installed capacity remains constant. This is due to the renewal of wind and solar farms. These values represent 2 to 4.5% of the annual energy generated by the infrastructure between 2010 and 2050 and 1.5% after 2050. For the Ecofys-WWF scenario, energy consumption increases from 63 to 950 TWh/year between 2000 and 2050 (between 0.4% and 3% of the produced energy) and stabilizes at 400 TWh/year after 2050. For the Bluemap scenario, the consumed energy increases to 250 TWh/year in 2050 and remains below % of the produced energy. These values omit the energy consumed to produce other raw materials in the infrastructure, as well as the energy consumed by the numerous industrial processes that transform raw materials into manufactured products, infrastructure construction, maintenance and repair operations. However, they indicate that the construction of infrastructures in the WWF and Garcia-Olivarès scenorios require much greater amounts of energy than the Bluemap scenario, which retains a fraction of the electricity produced from fossil fuels. The cumulative CO_2 emitted by electricity generation and by the production of infrastructure raw materials is shown in Figure 6.6. Even if the CO_2 intensity of raw materials production declines sharply between 2010 and 2050 (Table 6.1), the enormous increase in their consumption and the energy required to produce them is responsible for an increase in cumulative CO_2 emissions. Nevertheless, the high penetration of renewables in the WWF and Garcia-Olivarès scenarios allows the sum of emissions from electricity production plus those associated with the production of raw materials to be reduced. The balance sheet is therefore negative from a carbon balance point of view (emission reduction), which is obviously the foremost objective of these scenarios. Although it is less ambitious in terms of renewables, the Bluemap scenario makes it also possible to significantly reduce the sum of cumulative emissions compared to the historical trend (Figure 6.6, on the right). In 2050, the cumulative emitted CO_2 gain is about 230 billion tons. Throughout the deployment period, the sum of cumulative CO_2 emissions is similar for all three scenarios, as the gain in emissions from energy production is offset by over-consuming the energy required to produce raw materials. Differences only become noticeable after 2050 (Figure 6.6). We also see that the WWF and Garcia-Olivarès scenarios show similar trends in cumulative CO_2 emissions, although the amount of energy produced in 2050 by the WWF scenario (29,000 TWh/year) is much lower than that in the Garcia-Olivarès scenario (1,100,000 TWh/year). This is explained by the faster change of infrastructure and abandoning of fossil fuels in the Garcia-Olivarès scenario. This rapidity compensates for the emissions resulting from the production of raw materials, but the CO_2 balance alone should not mask the constraints of the Garcia-Olivarès scenario. This scenario involves a huge and very rapid consumption of raw materials and energy to renew the infrastructure (Figures 6.4 to 6.6). Obviously, such consumption would have environmental impacts other than those related to greenhouse gas emissions alone, and these would have to be quantified to assess the actual environmental gain.

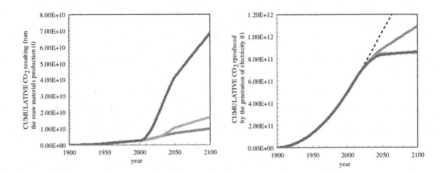

Figure 6.6. *Cumulative amount of CO_2 associated with the production of concrete, steel, aluminum and copper for the electricity-generation infrastructure (left) and cumulative amount emitted by the infrastructure + production of raw materials (right). Red line: Garcia-Olivarès et al. [GAR 12], green line: Ecofys-WWF [ECO 12] and blue line: Bluemap (IEA 2010), dotted black line: historical trend. For a color version of this figure, see www.iste.co.uk/vidal/energy.zip*

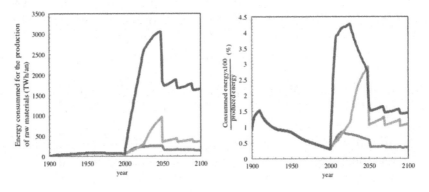

Figure 6.7. *Energy consumed in the production of concrete, steel, aluminum and copper in the infrastructure (left), and percentage of energy produced by the infrastructure (right). Red line: Garcia-Olivarès et al. [GAR 12], green line: Ecofys-WWF [ECO 12] and blue line: Bluemap (IEA 2010). For a color version of this figure, see www.iste.co.uk/vidal/energy.zip*

6.1.3. Comparison of raw material requirements with global consumption and production trends

In Figure 6.8, the annual steel, aluminum and copper requirements for the energy infrastructure are compared to global production. The share of annual global production required for the energy transition is reasonable for the Bluemap scenario (between 0.8% and 3% of global steel production, between 1.5% and 6% for aluminum and between 2% and 5% for copper). It is higher for the Ecofys-WWF scenario (up to 9.5% for steel, 15% for aluminum and 10% for copper between 2040 and 2050) and huge for Garcia-Olivarès (approximately 9% for aluminum and 25% for steel and copper between 2025 and 2050). For copper, the maximum consumption is reached at the peak of global primary production predicted by Northey *et al.* [NOR 14]. If this prediction is correct, implementation of the Ecofys-WWF and Garcia-Olivarès scenarios could be compromised by the availability of primary copper. In any case, a significant share of the world's copper production will have to be set aside for the production and use of electricity and/or copper substitutes will have to be found for these applications. The same trends are observed for the recycled share: between 20 and 30% steel, 10 to 20% aluminum and 30 to 50% of global recycled copper between 2025 and 2075 would be used for electricity-generation infrastructure according to the Garcia-Olivarès scenario. It is impossible to increase the share of recycled metal, which is not only limited by the recycling rate and average lifetime of manufactured products, but also by the amount of metals available in the end-of-life products. The share of scrap metal in total production (primary + recycled) is imposed by the quantity of primary production 30 years earlier. Between 2000 and 2030, for example, most recycled steel is primary steel produced between 1970 and 2000 (700 to 800 Mt/year). For the increase in overall steel demand shown in Figure 6.8, the share of scrap metal decreases from 38 to 22% between 2000 and 2010 and remains stable at 22% until 2030. This share then increases to 45% in 2050. The same trends can be seen for aluminum and copper, for which the primary production follows a similar pattern to that of steel since the 1970s. A substantial increase in recycling after 2030 is accompanied by an equally significant reduction in energy consumption for metal production. This widespread recycling will require a huge expansion of the present infrastructure of recycling in 20 years, in order to at least double the collection and management of end-of-life products and the flow of metals to be recycled.

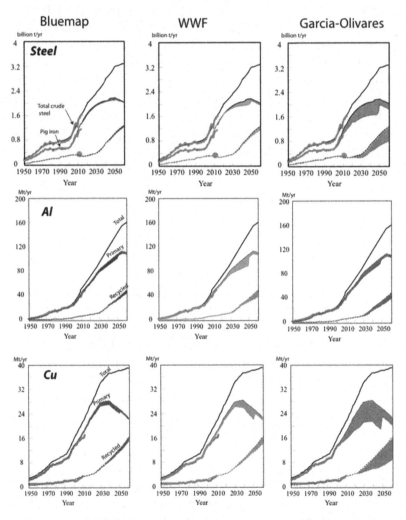

Figure 6.8. *Historical data (thick gray lines) and modeled data for global primary and secondary (recycled) production. The blue, green and red areas show the share of annual production to be reserved for energy-generation infrastructure in the Bluemap, Ecofys-WWF and Garcia-Olivarès scenarios. For a color version of this figure, see www.iste.co.uk/vidal/energy.zip*

The enormous demand for metals in the Garcia-Olivarès scenario is not only due to the transition from fossil fuels to renewable energies, but also to the huge global electricity production, which reaches 110,000 TWh/year in 2050. In comparison, electricity production in the Ecofys-WWF and Bluemap scenarios is 29,300 TWh/year and 36,659 TWh/year, respectively. Ecofys-WWF predicts a decrease in global energy consumption, which is expected to fall from 89,000 TWh/year in 2030 to 7, 000 TWh/year in 2050. Under these conditions, the share of global energy consumption used for cement, aluminum and steel production remains fairly constant in the Garcia-Olivarès scenario (from 13 to 15% between 2010 and 2050). On the other hand, this share increases in the Ecofys-WWF scenario from 13 to 22% over the same period, because the global energy consumption is decreasing while the consumption of raw materials is increasing. This 22% of global energy consumption would translate into about half the industrial energy consumption being reserved for the production of raw materials. In addition to the constraints on copper availability, the Ecofys-WWF scenario therefore raises the problem of energy availability for global production of cement, steel and aluminum (see Chapter 1). In order to maintain the proportion of global energy consumed by raw material production at 10 to 15%, global energy production must remain above 100,000 TWh/year in 2050, even if the efficiency of raw material production is improved and the metal recycling rate increases significantly. This value is 30% higher than the one foreseen in the Ecofys-WWF scenario (72,000 TWh/year in 2050), which is extraordinarily low compared to most published estimates for world energy consumption in 2050 (between 125,000 and 200,000 TWh/year).

6.1.4. Conclusion

Differing raw material and energy requirements have been estimated for different scenarios. These differences occur due to the contrasting energy mixes that are foreseen and the different installed capacities. The enormous amount of raw materials required by the Garcia-Olivarès *et al.* [GAR 12] scenario, which prohibits all forms of energy except renewable electricity, makes it unrealistic. It is limited by the required production of copper and other metals. The Ecofys-WWF scenario is less demanding, but the requirements are still high. By 2050, the cumulative primary copper content for the energy-generation infrastructure (approximately 53 Mt, see Figures 6.4 and 6.5) would be equivalent to three times the copper production of the Kennecot mine in Utah (USA) since it began being mined. This open-pit mine is 3.2 km long, 1.2 km wide and 1.2 km deep and is considered to be one of the largest man-made excavations on Earth (Figure 6.9). Since 1906, 6 billion tons of rock have been mined to produce 18 million tons of copper, which is equivalent to one year of current world production. Copper requirements for the only electricity-generation infrastructure in the Ecofys-WWF scenario would require producing in 30 years the equivalent of three times the Kennecot mine's production over a century, although the raw material requirements of the Ecofys-WWF scenario are minimal, as the installed capacity in 2050 is low.

Future technological innovations could help in reducing the amount of raw materials for renewable technologies, but the reverse is also possible (see Chapter 5). Moreover, the future of "old" infrastructure when more efficient technologies mature is uncertain. In Germany, 116 older but functional wind turbines were dismantled and replaced by 80 new and more powerful ones in 2010 (World Steel Association, 2012). Ideally, older turbines can be remanufactured and reused at other sites that require less capacity, which could extend their useful life. But if this is not the case, recycling before the end of operational life will increase the losses and especially the amount of energy used to build the energy-generating fleet. Similarly, our assumption that metals of the nuclear power plan can be recycled (Figure 6.2) does not correspond to the reality. In reality, only metals from the secondary circuit can be easily recycled. For all these and many other reasons, our estimates are subject to significant uncertainties. However, at least three conclusions can be drawn from the model:

– With today's technologies, the transition to low-carbon energy requires a substantial increase in consumption of structural raw materials. This is consistent with previous estimates from Kleijn *et al.* [KLE 12], Vidal *et al.* [VID 13] and studies on critical metals (Örhlund [ÖRH 12], Alonso *et al.* [ALO 12], Moss *et al.* [MOS 13]). It contrasts with the conclusions of [HER 15], which only considered the least demanding Bluemap scenario. The consumption of raw materials is highest during the deployment period, but remains significant after that period to renew the short life span of the infrastructure. In this context, it is important to identify the least expensive technologies before they are deployed on a large scale. The yield, financial cost and cost per KWh for the consumer are of course important, but the technical-economical optimum should not be the only criterion. Other factors should be taken into account, such as the supply of raw materials, recycling of installations and the environmental impact of building them. An analysis needs to be done for all raw materials and all areas affected by the energy transition; this should be extended to energy storage and distribution requirements. According to Garcia-Olivarès *et al.* [GAR 12], copper's share of the renewable energy infrastructure would barely account for 55% of the total amount required to shift from the current energy mix to a mix based on electricity generated from renewable energy. The cumulative copper values in Figure 6.4 are therefore significantly underestimated in relation to total requirements. Energy needs are also underestimated, partly because they only concern the infrastructure for generating electricity, but also because they only represent the energy needed to produce the raw materials. They omit all energy requirements for production of parts and machinery, transport, construction and maintenance operations, etc., which are necessary to manufacture and maintain a functional infrastructure.

– The needs in terms of raw materials and energy for the energy transition occur during a period of unprecedented demand in our history and rapid global growth, while the potential for recycling remains limited. The supply of metals from recycled sources is expected to increase, but will not exceed primary supply until the end of the scenarios. This is due to the delay between primary production and recycling, which limits the amount of metals available for recycling to that produced decades earlier. In addition, since recycling will never reach 100%, the cumulative amount of lost metals will increase over time and with the shorter lifetime of manufactured products. Assuming a lifetime of 20 years for renewable energy production facilities, 25% by weight of cumulative copper production since 2000 is lost by 2050 and 50% by 2100. These metal losses will have to be compensated by primary extraction.

– Structural raw materials such as concrete, steel, aluminum and copper are difficult to substitute, even though they require most of the energy used to produce mineral resources. These raw materials are used in large quantities to build urban centers that are developing all over the world. For these reasons, the supply of structural raw materials that do not yet have an apparent shortage is as important as the supply of so-called "critical" elements produced in much smaller quantities, for which the use is likely to change rapidly with technological innovation. For example, rare-earth elements are considered to be of critical importance for renewable energies, particularly for permanent magnet wind turbines containing neodymium, dysprosium and praseodymium. However, it is now possible to manufacture high-performance permanent magnet synchronous motors free of rare-earth metals. Similarly, electric motors that use electromagnets instead of permanent magnets are an option for future electric vehicles. It is therefore likely that the availability of scarce rare-earth element will not remain a problem for the transition to renewable energy in the long term, should primary supply become a problem. This is not the case for steel, concrete and copper, for which the availability is a prerequisite.

Figure 6.9. *Photograph of the Kennecot mine (Utah, USA). The white circle shows one of the huge trucks used in the open-pit mines (bottom photo)*

6.2. Dynamic model of the evolution of primary production-reserve-prices of mineral raw materials

6.2.1. *Resources, reserves and requirements*

In the early 1970s, Meadows *et al.* [MEA 72] suggested that many minerals would be exhausted in the short term. This has not been the case. On the contrary, annual production and primary reserve data published by major producing and consuming countries (like Australia, Canada, South Africa) show that despite ever-increasing utilization and production, reserves of base metals and their "life expectancy" or the remaining time before exhaustion of the known reserves have remained stable over the past 50 years. The life expectancy of most fossil fuel resources has been maintained because most reserves have grown at least as fast as production, and current reserves of many metals are larger now than they were 50 or 100 years ago. Figure 6.10 shows that USGS reserves of copper, iron, nickel and lithium have not decreased over the past 20 years, while production of these metals has doubled. This figure deserves some further comment: a deposit is a concentration for which the mineral nature and geometry make it possible to consider exploitation. This definition responds to the notion of a resource that evaluates the quantity of ore present *in situ*. It is a geological notion that is theoretically independent of time. A mining project is the transformation of a resource into a reserve and its economic exploitation. A reserve is therefore linked to an extraction plan, which is generally considered in the short term and for one of the economic conditions specific to the mined ore deposit, including the operating cost compared to anticipated revenues, which are naturally a function of the selling price of the mined ore. An important task in mining is to renew reserves, in other words to develop an extraction plan for the portion of resources that is not included in reserves at the present time. These resources correspond to the USGS "base reserves" shown in Figure 6.10. Compared to an initial mining operation plan, which is designed to pay back the capital invested as quickly as possible, sub-economic or low-return resources (accounted for in the "base reserves") can become economic ("reserves") simply because of the existence of an already-profitable infrastructure. When defining a deposit, the notion of reserve corresponds to the short term (financial equilibrium of the company), while the notion of resource and "base reserves" corresponds to the medium and long term. A reserve may decrease for several reasons: depletion due to exploitation of course, but also due to a decrease in the price of the metals[3], for regulatory reasons or because of a lack of other resources to extract them (energy and water, for example). In the first case, the corresponding resource decreases while it remains intact in the other cases. Conversely, the reserve may increase from a known resource as a result of technological innovations that allow metal to be extracted from an ore that was mined for other substances (for example, lithium through a process of extraction from lepidolite mica or cookeite chlorite, giving access to many resources that are currently untapped or unexploited for other substances). Technological innovation can also reduce production costs and extract metal at the same cost from less concentrated deposits. In this case, the reserve also increases because the volume of diluted desposits is larger than that of highly concentrated desposits. Copper deposits mined in the 1850s were commonly found to contain 5 wt% copper. They have been exhausted, but copper production has not stopped and the deflated price of copper has remained relatively stable over the long term. Copper is now produced from rocks that contain less than 1 wt% copper, which was inconceivable with the technologies used in 1850. As the volumes of deposits containing 1 wt% copper are much larger than those

---

3 A recent example of possible loss in reserve value occurred in January 2015, when the value of reserves at the world's top 10 copper mines plunged by $130 billion in two days: https://mlms.infomine.com/ga/Mining%20News%20Digest/20150114/Link /http/www.mining.com/top-10-copper-mines-plunge-134-billion-in-value-66653/?utm _source=digest-en-mining-150114&utm_medium=email&utm_campaign=digest&g ae=945843

containing 5 wt%, the share of exploitable copper has increased over time. Obviously, reserves also increase with the discovery of new deposits... which are only discovered if they are searched for. As long as the mining industry can exploit known deposits at a profit, exploration to find new deposits is not favored because it is expensive. It follows that the integrality of future copper reserves and resources are not currently known. There are certainly buried or scarcely accessible exploitable resources that will be discovered in the future, when current exploited deposits will no longer be sufficient and exploration and exploitation technologies will have progressed. This optimistic view contrasts with the pessimistic one in many studies, which insist on the finite nature of the Earth's mineral resources and the impossibility of continuous exponential growth. If continued exponential growth in copper production could be sustained at a rate of 3%/year (rate observed over the past century), all the copper in the Earth's crust (30 km thick, with an average concentration of 0.001 wt%) would be mined in 620 years to form a continuous 30-meter thick layer covering the entire Earth. This assumption of continued growth is absurd, not only because we will probably never have the capacity to extract all the copper in the Earth's crust, but also because humanity will never need it. The current growth in metal production can only be a strawfire at the scale of humanity and will not be able to continue beyond a few decades. The question is not how long we can continue the exponential evolution, but rather whether we can extract the mineral resources required for growth of the needs identified over a period of 50 to 100 years at reasonable cost and environmental impacts. Beyond this, future needs, technologies and sources are so uncertain that the question makes little sense. In the following, we test different demand growth scenarios for the next 50 to 100 years and try to model the coupling between physical and monetary flows.

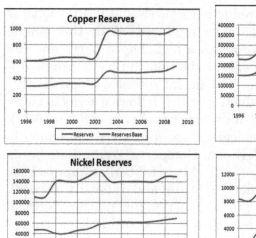

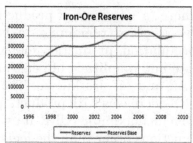

Figure 6.10. *Changes in reserves of copper, iron, nickel and lithium between 1996 and 2010 (source: USGS). According to the USGS nomenclature, "reserves base" is the share of an identified resource that meets minimum physical and chemical mining and production criteria in accordance with current practices, including concentration, quality, thickness and depth. The "reserves base" includes the "reserves" and portion of resources that have a reasonable potential to become economically available in planning horizons beyond those based on current technology and economics. For a color version of this figure, see www.iste.co.uk/vidal/energy.zip*

6.2.2. Modeling the supply and primary reserves of fossil fuel resources

6.2.2.1. Limitations of models that assume a static stock of ultimate recoverable resources (URR)

Hubbert [HUB 56, HUB 62] suggested that conventional oil production in the United States was based on a finite stock of "ultimate recoverable resource" (URR), which is itself proportional to cumulative oil production over time (Q(t)). Production over time follows a logistic curve:

$$dQ/dt = a \bullet Q(t) \bullet [URR - Q(t)]$$

where "a" is a rate of production growth and [URR - Q(t)] is the resource remaining to be produced. With this simple formulation, production follows a bell-shaped curve and the date and amplitude of peak production can be determined if URR and "a" are known. The initial resource stock, [URR - Q(t)] decreases with production which follows a sigmoid curve from its maximum (URR at t = 0) to zero at t = ∞ (Figure 6.11). Functions other than the logistic equation were subsequently proposed, such as the Verhulst, Gompertz, Weibull, Richards and Johnson functions [FRI 11]. Hubbert's approach is generally used to predict the date and intensity of peak production by calibrating the production function on historical data and assuming the "ultimate recoverable resource" (URR). In reality, the historical production of fossil fuel resources does not follow a perfect logistic function because production is stimulated by demand and demand can change over time. Between 1970 and 2000, the growth rate of base metal production was lower than from the 1950s to 1970s and since the 2000s (Figure 1.1). This lower rate of production growth could have been mistakenly interpreted as a sign of depletion of reserves. It was in fact the result of a decrease in demand. The sudden increase in demand at the beginning of the 21st Century, spurred by China's rapid development, was met with an increase in production without any long-term supply restriction. Mining history also shows that the types of mined deposits change radically over time, and technologies evolve rapidly. Modeling historical production series by a single sigmoid curve for a given commodity thus represents a gross simplification of reality, and a logistic projection made at the beginning of the Bronze Age would have predicted the depletion of copper sources in the short term. To overcome this problem, some authors have proposed to model the same series of historical data with several logistic curves (for example, [SVE 14a]). This approach does not provide any additional information on the driver of demand variation, and while it is possible to reproduce historical production variations more accurately, the prediction of future variations remains uncertain. Actually, a problem with Hubbert-type logistic approaches is their deterministic nature, which implies that demand is growing at a constant rate and that production follows this trend as long as the predefined size of the URR stock allows. Even under the assumption that demand is growing at constant rate, it is very difficult to value the URR reserve stock. In theory, this stock can be estimated from geological data or it can be derived from historical production data using the Hubbert linearization method [HUB 82]. In the case of copper, highly contrasting URR values have been estimated using the same historical data series. These range from 1.5 Gt [LAH 10] to 3.8 Gt [FRI 11]. This enormous difference lies in the procedure used to calibrate a single logistic function of a given shape on scattered historical data. Frimmel and Müller [FRI 11] showed that copper production for the period 1900 to 2010 could be approached by a logistic function with URRs as different as 1.9 or 3.8 Gt Cu. If the historical data were to show the regular distribution of a logistic curve, the URR could have been estimated from any limited series of production data. However, Hubbert's linearization method for 1970 to 1995 production indicates a URR of 1 Gt, while the trend from 2000 to 2010 indicates 1.5 Gt [LAH 10]. Therefore, the URRs estimated from production data are not the actual ultimate reserves. In 2010, the identified copper reserves were 1.1 Gt (USGS data), but three years later, the USGS reported that approximately 3.5 Gt of undiscovered copper was to be added [JOH 14]. These data show that the uncertainties on copper URRs estimated from geological data or historical production data are enormous. The uncertainties of future trends modeled with a fixed stock of initial resources (URR) and a logistic approach are equally enormous. Although there is little doubt that the production of a fossil substance will eventually decrease after increasing, the predictive power for the next 50 years of classical Hubbert-type approaches is very limited. Despite their evident limitations, the simplicity of these approaches still has a strong influence on the modeling of fossil fuel resource production, and several economists have attempted to reconcile economic theory with the bell curves obtained with Hubbert's approach (see Okullo *et al.* [OKU 15], Reynès *et al.* [REY 10]).

6.2.2.2. *Dynamic prey–predator approach*

The same bell-shaped production curves are also obtained with non-empirical prey–predator models. The prey–predator model was originally proposed by Lotka [LOT 25] and Volterra [VOL 26] to describe the dynamics of competition in simple biological systems between two animal species such as wolves (W) and rabbits (R):

$$dR/dt = R(t)*\alpha - W(t)*R(t)*\beta \qquad\qquad [6.1]$$

$$dW/dt = W(t)*R(t)*\delta - W(t)*\gamma \qquad\qquad [6.2]$$

where α is the birth rate of rabbits and γ is the mortality rate of wolves, β is the predation coefficient of wolves on rabbits and δ is the efficiency with which an eaten rabbit is transformed into an unborn wolf. For constant values of α, β, δ and γ, the equations have periodic and time-shifted solutions (Figure 6.11b). Prey–predator models were introduced into the economy by Goodwin [GOO 67] to replicate endogenous cycles of economic activity. Brander and Taylor [BRA 98] used equations similar to [6.1] and [6.2] to model the cycles of human (predator) abundance and famine that depend on the availability and production of renewable food resources (prey) in closed systems such as on Easter Island. These authors showed that cycles can result from overexploitation of resources, leading to the collapse of society. Their Ricardo-Malthus model was a source of numerous derived articles that examined the influence of various economic factors or the stratification of society ([DAL 00, DAL 05, DEC 05, REU 00, PEZ 03, MAX 00, MOT 14] and others). Bardi and Lavacchi [BAR 09] examined various situations in which the production of a natural resource (prey) depends on the stock of capital used in its

production (predator) (Figure 6.11c). The capital stock is
defined as the overall amount of all economic resources used
in exploitation. Bardi and Lavacchi [BAR 09] provided
examples where the nature of capital stock was as different as
the number of miners employed in gold mining, the number of
vessels used in whaling, or the amount of exploratory drilling
in the oil industry. In all these examples, the model generates
a "Hubbert curve" with output growth followed by a decline as
the stock of available resources is depleted. However, contrary
to Hubbert's empirical approach, the reasons for growth and
decline are explicit here. The dynamics of the system are
controled by two feedback loops, one positive resulting from
the reinvestment of benefits generated by the resource and
the other negative resulting from the progressive depletion of
easily accessible and inexpensive resources. Another
characteristic that the Bardi and Lavacchi and Hubbert
approaches both have in common is that the initial resource
stock must be known (number of prey at t0 = URR), as the
rate of regeneration of fossil fuel resources is considered to be
zero. As we have seen above, it is problematic to use static
URR because the stock of exploitable resources at t is not
known at time t0. The only known value is the historical
evolution of reserves over time. In the case of copper, reserves
have increased from about 25 Mt in 1900 to about 700 Mt
today, at a rate proportional to production. Of course, this
increase in reserves is not a regeneration in the sense of
renewable resources such as rabbits or wolves, but it can be
modeled as such through the first term on the right of
equation [6.1]. If future demand increases faster than
"regeneration" of reserves, production becomes constrained by
the rate of reserve growth (flow) rather than by the size of the
URR (stock). The rate of reserve growth is therefore an
important parameter that needs to be incorporated into the
model.

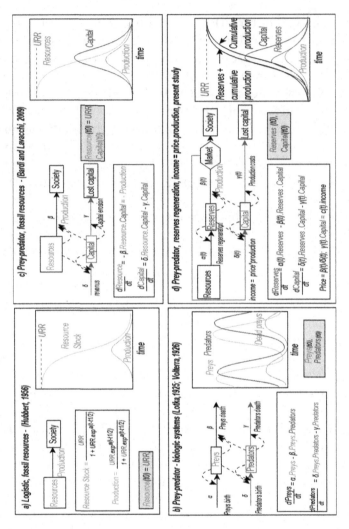

Figure 6.11. *Comparison of different approaches used to model historic trends and future resource production (a, c, d) or to model changes in prey and predator populations in biological systems (b). The boxes are the stocks and the solid arrows are the flows, the dotted arrows show the coupling of equations. The initial conditions are listed in the gray boxes. Contrary to (c), the actual stock of reserves in (d) increases over time, with regeneration rate α. Changes in capital are controled by annual revenues from the mining industry (δ(t).capital.reserves) and the annual cost of production (γ(t).capital), where δ is the efficiency with which the resource is transformed into capital and γ is the rate of capital erosion. For a color version of this figure, see www.iste.co.uk/vidal/energy.zip*

6.2.2.3. *Regeneration of reserves*

Unlike renewable resources with a fixed reproductive rate, α is allowed to change over time. In particular, it can be equal to zero, in which case regeneration stops regardless of the size of known reserves and capital. Regeneration rate α depends on many technical, geological, economic, financial, social, environmental and geopolitical factors such as improved geological knowledge and exploration investments that allow for new discoveries, as well as generally improved efficiency of mining operations that lower production costs and enable economically viable extraction of metal from low-grade deposits. This point is illustrated by the log-linear relationship between concentration (C_t %) and cumulative tonnage produced $(CTCu)_t$, as observed by many authors (see [GER 08, VIE 12]):

$$\log(CTCu)_t = (C_t \% - a)/b \qquad [6.3]$$

where a = 2.064, b = -0.237 for sulphur porphyry [GER 08].

The change in cumulative production with a decrease in concentration is illustrated for copper in Figure 6.12, assuming that all copper is produced by porphyries from 1900 to the present day. This assumption does not correspond to reality, but we see that a drop in ore grade from 2% in 1900 to 0.6% in 2005 allows us to reproduce the cumulative amount of copper produced worldwide over the same period. In other words, a simple drop in concentration of exploited rocks makes it possible to obtain a cumulative production that evolves exponentially, as has been seen since 1900. It is important to understand that the evolution shown in Figure 6.10 assumes that the *same deposits* with a unimodal Gaussian distribution of copper concentrations are mined, starting with the most concentrated parts and going towards increasingly less concentrated parts when the most concentrated parts have been exhausted. An important conclusion from the product-concentration cumulative tonnage curves is that reserves increase exponentially with declining grade of deposits, even in the absence of new discoveries. This presupposes, of course, that improved technologies will allow less and less concentrated rocks to be exploited while maintaining a relatively stable cost of production. It remains difficult to provide a non-empirical equation for "regeneration" of reserves that captures all the above-mentioned parameters, some of which are economic, social and geopolitical. However, a high value of α in equation [6.1] (high level of regeneration) is expected at the beginning of fossil fuel resource development, when the discovery of concentrated and accessible deposits is relatively easy. Equation [6.1] indicates that reserves are constant over time (dR/dt = 0) for α = W*β = production/reserve. Since 1900, α has been higher than the production/reserve ratio, but the difference between α and production/reserve has been decreasing over time as the ratio of production/reserve has been increasing since 1950. When the cumulative amount of produced copper approaches the unknown URR, regeneration will inevitably become zero. This suggests that α should decrease on average over time. As long as α remains above the production/reserve ratio, reserves will continue to grow, they will peak at α = production/reserve and collapse when α is below the production/reserve ratio.

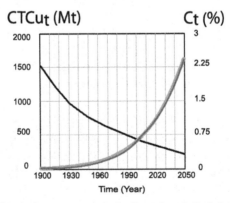

Figure 6.12. *Estimated copper concentration of exploited deposits (C$_t$%, black curve) to replicate observed global cumulative production (CTCu$_t$, blue curve) with equation 15 (red curve). For a color version of this figure, see www.iste.co.uk/vidal/energy.zip*

6.2.3. Modified prey–predator model

The predator stock W(t) in equations [6.1] and [6.2] corresponds to the Wealth or capital of the mining industry and the stock of prey R(t) corresponds to copper Reserves (Figure 6.11d). W(t) is the aggregation of economic resources used to produce primary copper. It encompasses industrial infrastructure and all other forms of capital in the mining and processing industry, as well as information and know-how. The annual copper production in Mt/year is given by the second term of equation [6.1] (W(t)*R(t)*ß) where ß is the productivity of capital, i.e. the efficiency with which capital is used for copper production. The evolution of W(t) over time is given by equation [6.2], where the first term represents the annual revenues of the mining industry (predator births) and the second term represents the cost of production (predator deaths). Revenues are proportional to δ (Mt*year)$^{-1}$, which describes the transformation efficiency of copper into wealth. Revenues are also given by the amount of sold copper (Mt/year) multiplied by the price of copper ($/ton), such that:

$$W(t)*R(t)*\beta(t)*price(t) = W(t)*R(t)*\delta(t) \qquad [6.4]$$

Equation [6.4] can be rearranged to express δ(t) in terms of price (in constant 1998 dollars, i.e. adjusted for inflation) and ß(t):

$$\delta(t) = price(t)*\beta(t) \qquad [6.5]$$

The cost of copper production (US$$_{1998}$/t) includes all mining costs, depreciation and amortization, corporate overhead, royalties and other financial interest. Historical production costs can be estimated as a fraction c(t) of income, and the term W(t)*γ(t) in equation [6.2] is then replaced by c(t)*revenues:

$$dW/dt = W(t)*R(t)*price(t)*\beta(t)*(1-c(t)) \qquad [6.6]$$

In the following, equations [6.1] to [6.6] are used to estimate the parameters α(t), ß(t), δ(t) and c(t) and their variations from t0 = 1900 to t = 2015. These parameters are constrained by historical data on copper industry prices, production, reserves, production costs and revenues. Using equations [6.1], [6.6] and the equations below, the future of production can then be modeled, as well as reserves and capital for different evolution scenarios for $\alpha(t)$, $\beta(t)$, price(t), c(t) and $\gamma(t)$, which are compatible with historical trends:

$$W(t)*\gamma(t) = c(t)*revenues(t), \text{ thus} \qquad [6.7]$$

$$\gamma(t) = price(t).\beta(t).R(t).c(t)x \qquad [6.8]$$

Equations [6.7] and [6.8] are used to estimate the future price of copper for specific trends of $\gamma(t)$ or c(t) and demand.

6.2.4. *Model constraints, initial conditions and historical data modeling*

	2014	2013	2012	2011	2010	2009	2008	2007	2006	2005	2004	2003	2002
Deflated Price[1] (US$$_{1998}$/t)	5 240	5 240	5 750	6 490	5 740	4 040	5 330	5 690	5 610	3 190	2 550	1 670	1 510
Real Price[1] (US$/t)	7 490	7 490	8 100	8 950	7 680	5 320	7 040	7 230	6 940	3 830	2 950	1 880	1670
All metals produced by the 40 « majors »													
Total revenues[2] (bUS$)	453	482	525	539	435	325	349	312	249	222	184	110	93
Revenus[5] (bUS$$_{1998}$)	317	337	373	391	325	247	264	246	201	185	159	98	84
Net profits[2] (bUS$)	45	21	66	132	110	49	57	80	66	45	28	12	6
net profits[6] (bUS$$_{1998}$)	33	15	48	96	82	37	43	63	53	37	24	11	5
Margin[3] (%)	10	4	13	24	25	15	16	26	27	20	15	11	6
c(t)[4] (all metals)	0.9	0.96	0.87	0.76	0.75	0.85	0.84	0.74	0.73	0.80	0.85	0.89	0.94
Copper													
Revenus[7] (bUS$$_{1998}$)	91	96	106	112	93	71	75	70	58	53	45	28	24
Net profits[8] (bUS$$_{1998}$)	9	4	13	27	23	11	12	18	15	11	7	3	2
Global production[1] (Mt)	17.0	17.0	16.9	16.3	16.0	16.0	15.6	15.5	15.0	15.0	14.5	14.0	13.5
Profits[9] (US$$_{1998}$/t Cu)	529	247	792	1678	1468	664	790	1161	1016	714	477	218	115
Profits[10] (US$/t)	756	353	1116	2314	1964	875	1044	1475	1257	857	552	245	127
Production costs[11] (US$$_{1998}$ /t Cu)	4719	5012	5027	4901	4289	3431	4459	4231	4123	2543	2162	1488	1413
Production costs[12] (US$ /t Cu)	6746	7164	7082	6758	5738	4518	5890	5376	5100	3054	2501	1675	1562
c(t)[13] (copper)	**0.89**	**0.95**	**0.86**	**0.74**	**0.74**	**0.84**	**0.85**	**0.80**	**0.82**	**0.78**	**0.81**	**0.87**	**0.92**

Table 6.2. *Copper industry revenues, profits and production costs used to constrain the model. 1) USGS, 2) PwC (2003-2015), 3) Net profits*100/Revenues, 4) 1 -((5)-(Net Profits)/(5)/100)/100, 5&6) Total revenues/(Real Price/Delated Price), 7) (5)*0.2/0.70, 8) (6)*0.2/0.70, 9) (8)*1000/(Global production), 10) (9)*/(Real Price/Delated Price), 11) (Price in US$$_{1998}$ – (9)), 12) (11)*/(Real Price/Delated Price), 13) (11)/Price in US$$_{1998}$*

The historic evolution of $\alpha(t)$, $\beta(t)$, $\delta(t)$ and $c(t)$ since 1900 has been constrained in order to reproduce the change in reserves, global primary copper production, US$_{1998}$ prices from 1900 to 2015, and annual production costs from 2002 to 2013 (Figure 6.14 and Table 6.2). The 2002-2014 total revenues of the copper mining sector were calculated from the total revenues of the 40 major mining industries that produced 70% of the world's raw copper (PwC)[4]. Net revenues and profits of world raw copper production were estimated by multiplying the total revenues and net profits by the share of copper revenues (20%) divided by 70% (Table 6.2). We estimate that the value of $c(t)$ = production cost/income for the period 2002-2014 ranges from 0.73 to 0.96 and that production costs range from 1,400 to 5,000 US$_{1998}$/t. The significant increase in production costs from 2004 onwards results from the significant investments required to increase the production capacity. Due to the lack of data, $c(t)$ is assumed to drop from 0.85 in 1900 to 0.7 in 1935, thanks to improved technology. It remained stable at 0.7 until 1985, before reaching the values in Table 6.2 in 2002-2014. Reserves in 1900 were assumed to be R(1900) = 25 Mt [SCH 10]. The value of capital in 1900 is not documented, but a trial-and-error approach indicates that it should range between 250.10^9 and 660.10^9 US$_{1998}$. The calculations were made assuming that W(1900) = 600.10^9 US$_{1998}$.

The large variations in short-term production are controled by similar variations in the rate of capital productivity $\beta(t)$ (Figure 6.13). If demand declines, the industry can reduce its production at constant capital and reserves (decrease of $\beta(t)$) for a short period of time. However, a reduction in production leads to a reduction in revenues and profits, and a reduction in exploration investments. We therefore expect that the regeneration rate of reserve follows the same variations as $\beta(t)$. The model was constrained for $\alpha(t) = \beta(t)/M$ with M

---

4 PwC: PricewaterhouseCoopers LLP, Annual Review of Global Trends in the Mining Industry – Mine 2003 to 2015, http://www.pwc.com/mining.

estimated to reproduce historical changes in reserves and production, $\delta(t) = \beta(t)/price$,   and $c(t) = production$ costs/revenues. The estimated variations of $\alpha(t)$, $\beta(t)$, $\delta(t)$ and $c(t)$ are shown in Figure 6.13.

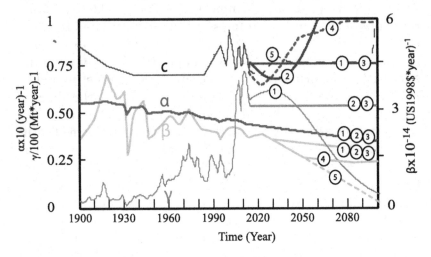

Figure 6.13. *Evolution of alpha, beta, delta, c and gamma. Evolution after 2015 correspond to scenarios 1 to 5 discussed in the text. For a color version of this figure, see www.iste.co.uk/vidal/energy.zip*

6.2.5. Long-term future trends

Extrapolation of the model into the future requires additional assumptions regarding the evolution of parameters $\alpha(t)$, $\beta(t)$, $\delta(t)$ and $c(t)$. We examine different scenarios in the following, which foresee various changes in demand, $c(t) = revenues/production$ costs, and prices. These different scenarios are not intended to accurately predict the evolution of copper production, which is impossible, but they do give an insight into the nature of the linkages between flows and stocks of money and material. They also provide information on the system dynamics and the production conditions required to reach stability.

6.2.5.1. *Business as usual scenario leading to a peak in production*

Scenario 1: constant c(t) = costs of production/revenues and prices, change in capital erosion rate γ(t)

Assuming future variations of α and β are in line with the historical trends (Figure 6.13), scenario 1 at constant prices and c(t) leads to a peak in copper production in 2040 (Figure 6.14). The peak production at 37 Mt/year is intermediate between the demand predicted by Northey *et al.* [NOR 14] and Meinert *et al.* [MEI 16] (26-30 Mt/year in 2050), and those estimated by Ayres *et al.* [AYR 03] (42-55 Mt/year). The date and size of the peak are similar to those obtained by Frimmel and Müller [FRI 11] with a URR of 3.8 Gt and Northey *et al.* [NOR 14] with a URR of 3,150 Mt. Peak production follows a peak in reserves reached 15 years earlier, when consumption exceeded regeneration. Unlike production and reserves, capital does not collapse immediately. It continues to increase after the reserves and production peaks. This evolution is constrained by the assumption that c(t) = 0.75. A c(t) value of less than 1 implies that the calculated production costs are always lower than the revenues, so profits increase regardless of the level of production. However, keeping the production costs below the revenues at constant price is unrealistic during a period of declining production (after the peak). This would imply that the rate of capital erosion (γ(t)) decreases sharply over time (green curve 1 in Figure 6.13). In reality, γ(t) is likely to remain constant at best, implying that the wealth stock is likely to decline with production at constant prices (next scenario).

Scenario 2: γ and constant prices, change in c(t)

If γ(t) instead of c(t) is kept constant from 2015 onwards, c(t) at constant price increases rapidly and becomes > 1 after the production peak. At this stage, the production costs become more significant than the income and the capital W(t) decreases. This evolution implies that the industry "consumes" its wealth; it sells its infrastructure and reduces its number of employees because revenues are not high enough to cover expenses. Such a development naturally favors the collapse of production and leads to bankruptcy. The only way to prevent c(t) from becoming > 1 is to increase the price of copper (next scenario).

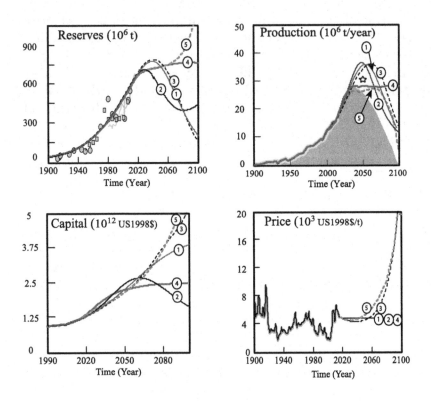

Figure 6.14. *Changes in reserves, production, industrial capital and copper price for the various scenarios 1 to 5. Historical data are shown in yellow, and model results in black. For a color version of this figure, see www.iste.co.uk/vidal/energy.zip*

Scenario 3: constant γ and c(t), price increase

If copper demand remains constant or increases after the peak production, the price of copper is also expected to increase as supply becomes smaller than demand. This price increase at constant γ(t) is accompanied by an increase in annual revenue (= production*price), which may remain higher than annual production costs. Contrary to scenario 2, c(t) in scenario 3 can be maintained < 1 and the wealth (capital) stock can continue to grow after the peak production. This capital growth helps to maintain production at its maximum level. This situation is illustrated in Figure 6.14, which shows the evolution of capital and price to maintain c(t) = 0.75 from 2015 to 2100.

In scenario 3, the increase in copper prices postpone the peak date and collapse of production by about 20 years. As long as the industry maintains its profits, investment in exploration and technology improvements is possible, which favors the regeneration of reserve when it becomes limited compared to the rate of production growth. This last point has not been considered in scenario 3 because it requires a coupling between profits and α(t) that could not be constrained with the available historic data. However, the possible feedback of profit onto regeneration should further postpone the peak production.

6.2.5.2. *Regeneration of reserves and/or demand decoupled from historic variations*

Unlike previous scenarios, future demand and production are assumed to stabilize or increase at a slower pace than in the past. This situation is modeled with β decreasing faster than the historical trend. The decline of β below historical trend implies that production (=β.Capital.Reserves) after 2015 is lower than the maximum possible production for existing reserves and capital stocks. This situation can have three different causes:

1) reduction of primary copper demand when the contribution of recycled copper becomes significant (see previous chapter) and when the level of copper saturation in society is approached (see Chapter 1). In this case, the price of copper may remain constant or even decrease;

2) inability of the industry to meet demand due to a lack of other resources (water, energy, etc.) or the decision to keep production below demand. In this case, the price of copper is expected to rise;

3) decoupling between production, reserves and capital for metals other than copper such as In, Ge, Co and other metals extracted as co-products of "major" or "carrier" metals such as Fe, Al, Mg, Ti, Sn, Ni, Cu, Pb, Zn, Cr and Mn. If the demand (and price) for a co-produced metal increases, large quantities cannot be extracted because volumes are determined by the extraction of major metals. For such commodities, an increase in production cannot meet the growing demand.

Scenario 4: stabilization of future demand at constant price

This scenario assumes that demand stops growing exponentially and stabilizes at 28 Mt/year from 2040 to 2100. Production was modeled with change in ẞ shown in Figure 6.13 at constant price and regeneration of reserves in line with historical trends. Contrary to previous cases, scenario 4 suggests that production, reserves and capital do not peak but stabilize up to 2100 (Figure 6.14). Stable primary production of about 30 Mt/year of copper at constant price from 2040 to 2100 is therefore possible.

Scenario 5: decline in constant demand production and price increase

A decrease in ẞ and production could also result from various factors that limit the production capacity, despite that the necessary reserves and capital exist. As mentioned above, this is generally the case for metals that are co-produced from "major" metals. In the case of a major metal such as copper, factors that limit the production capacity include the industry's decision to limit investment in new production capacity, geopolitical or environmental constraints, sustainable strikes, lack of other resources needed for exploitation, etc. If the global demand remains above the supply capacity, the price of copper is expected to rise. This is illustrated by curve 5 in Figure 6.14, which was calculated such that the price of copper increases so as to maintain $c(t) = 0.75$ for as long as possible. Production in scenario 5 stabilizes at about 27 Mt/year for 40 years, between 2030 and 2070. Of course, different price trajectories are possible and the production plateau can last longer for different evolutions of ß and $c(t)$. The important information in scenario 5 is that the rapid collapse of production after a sharp peak is not inevitable; peak production can be avoided or can at least be delayed by a few decades if demand stabilizes and prices can rise from 5000 to about 16 000 US$/t between 2020 and 2080.

6.2.6. Discussion and further extension of the prey–predator model

The results presented above show that if $\alpha(t)$, $\beta(t)$, $\delta(t)$, $\gamma(t)$ and $c(t)$ change in line with historical trends, production follows a bell curve similar to the results of conventional Hubbert approaches. However, a future decline in the growth rate of demand may lead to plateaus instead of peaks, or delay the peak date to beyond 2050 (scenarios 4 and 5). Moreover, even if production stabilizes or declines after 2050, this does not mean that there will be a global shortage of copper, as primary copper produced today is not lost and constitutes the reserves for future recycling. If world copper demand stabilizes at about 30 Mt/year from 2030 onwards (scenarios 4 and 5), recycling is expected to compensate for the possible decline in primary production after 2050 and thus avoid the collapse of supply (Figure 6.8).

The results presented above assume a capital stock in 1900 of 600.10^9 US\$$_{1998}$. Unfortunately, it has not been possible to find reliable constraints on this value and other initial conditions can replicate historical production and reserve data for the observed prices. Additional constraints can be provided by rationalizing the interdependencies between the model's parameters and starting conditions with the price for a series of different metals. Preliminary results show that if the historical data for different metals are modeled with constant values of β and α, capital and reserves in 1900 are inversely proportional to the average price and dilution of mined metals. They also suggest that β, β/α and c(t) are all proportional to an average price calculated for period 1950-2015. This is illustrated in Figure 6.15, which shows the evolution of production, reserves and cumulative production + reserves of Cu, Ni, Ag and Au.

A detailed comparison of modeling results for different fossil fuel resources should therefore help to better understand the interdependencies of the prey-predator model parameters and their relationship to price. The model results in Figure 6.15 also show that the production and "URR" (= cumulative production + reserves in 2100) are consistent with those obtained by Sverdrup and Ragnasdottir [SVE 14a] for copper and gold.

The situation is clearly different for silver and nickel. Nickel production does not decrease until 2070 in our case, with a URR estimated at about 600 Mt in 2100. On the contrary, Sverdrup and Ragnasdottir [SVE 14a] estimated a production peak in 2020 and a URR of 170 Mt. Assuming that future "regeneration" rates of nickel and silver reserves are the same as those observed since 1900, there is no clear reason why their production should collapse before 2060. In addition to a more systematic study to understand the relationship of the model's parameters and initial reserve and capital conditions with price, other extensions are being considered. Obviously, a price dependency on demand and costs would have to be introduced. Incorporating prey–predator dynamics into a macroeconomic model would be one way to understand which of the different scenarios considered above is more likely. The two equations [6.1] and [6.2] are very crude in that they do not explicitly describe the constraints imposed by the additional resources, which are crucial for copper extraction – particularly energy and water (scenario 5).

Several authors have suggested that the observed decrease in concentration of exploited deposits over time could lead to an exponential increase in the energy required to mine metal, which imposes a limit on production. However, we saw in Chapter 3 that improvements in energy efficiency over time have so far managed to offset the effect of declining concentrations. Future development will not only depend on the ore grade of exploited deposits, but also on the improvement of industrial processes and the share of renewable energy sources, global share of recycling, etc. All these parameters and their coupling could be inserted into a single model, but the uncertainties of their evolution prevent yet a single and robust estimate of future trends. It seems therefore more reasonable to estimate a range of possible trends constrained by the historical data instead of one single solution.

Finally, recycling must be introduced into the model because it is one of the keys to securing future supply. Recycling dynamics are expected to be similar to those reported for primary production, with a certain quantity of recycled copper being controled by secondary reserves and therefore available from end-of-life products. This is illustrated in the next chapter.

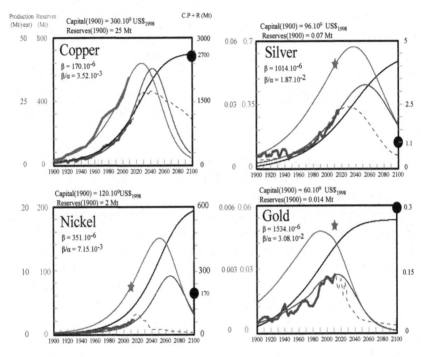

Figure 6.15. *Changes in copper, silver, nickel and gold reserves (red), production (blue) and cumulative production + reserves (black), assuming constant values of α, β and c, at constant price from 2015 onwards. The thick red line for copper and the red stars for other metals show known reserves. The thick blue lines show historical data of primary production and the dotted lines show the evolution modeled by Sverdrup and Ragnasdottir [SVE 14a]. The black dots show the URR values used by Sverdrup and Ragnasdottir [SVE 14a], which are compared with the values of cumulative production + reserves calculated with the prey-predator model (black line). Capital(1900) and Reserves(1900) decrease and β and β/α increase with the price of metal and its dilution in the mined ore deposits. For a color version of this figure, see www.iste.co.uk/vidal/energy.zip*

6.2.7. *Incorporation of recycling in dynamic models*

Ali *et al.* [ALI 17] (supplementary information S1) showed that the evolution of recycling over time can also be demonstrated by prey-predator-type dynamics. The complete model (Figure 6.16) is similar to the one shown in Figure 6.3, with an extension that integrates two markets: one for primary metals (Market I) and the other for secondary metals (Market II). As shown in Figure 6.11d, the market allows for cash flow (revenue) to be created from a material flow. Cash flows feed two capital stocks: the primary sector and recycling. Metal reserves to be recycled (EOL on Figure 6.16) are made up of accessible end-of-life products. They are supplied by two material flows, which are the primary and secondary flows produced 20 to 25 years earlier, in other words the average lifetime of consumer goods [GLÖ 13] multiplied by a rate R. This rate corresponds to the fraction of primary and secondary production that will be recycled, the remaining fraction being lost.

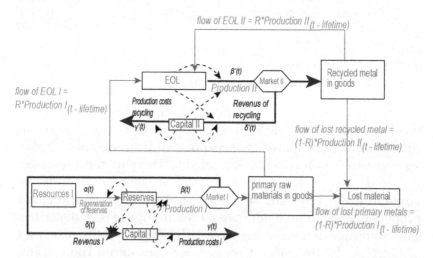

Figure 6.16. *Dynamic model with prey–predator coupling between flows and stock of material (green) and cash (blue). The boxes are stocks and the solid arrows are flows. Parameters α(t), β(t), δ(t), γ(t) control the primary sector and parameters β'(t), δ'(t) = price/β'(t), γ'(t) = γ(t)/4 control the secondary sector (recyling). For a color version of this figure, see www.iste.co.uk/vidal/energy.zip*

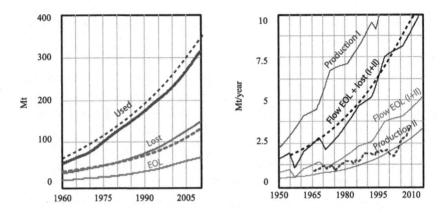

Figure 6.17. *Left: stocks of copper used (in consumer goods), lost and in the stock of accessible end-of-life products (EOL). Right: primary and secondary production flow towards EOL stock and lost EOL + stock. The solid curves are the results of the model in Figure 6.16 and the dotted curves are historical data from Glöser et al. [GLÖ 13]. For a color version of this figure, see www.iste.co.uk/vidal/energy.zip*

The quantity of metals available in end-of-life products (flow of EOL I and II, EOL stock) will increase sharply in the future, as the quantity incorporated into consumer goods has grown exponentially by about 3%/year for the past hundred years. However, increased recycling will only be possible if the recycling infrastructure (Capital II) increases proportionately. The model presented in Figure 6.16 describes the couplings between metal availability, capital size and production. In the case of copper, we considered the erosion rate of Capital II ($\gamma'(t)$) to be 4 times lower than Capital I ($\gamma(t)$) because the energy and production costs of recycled copper are lower than those of primary copper (see Chapter 3). With this assumption, historical data on used and lost copper stock, annual copper flow generated in end-of-life products and recycled copper flow (Figure 6.17) can be replicated for $\beta'(t) = 6.10^{-14}$ (US\$$_{1998}$.Mt)$^{-1}$, $\delta'(t) = $ price/$\beta'(t)$, R = 0.5 for an average lifetime for copper of 25 years.

The results of dynamic modeling up to the year 2100 (including recycling) are presented in Figure 6.18 for the two scenarios 3 and 4, which were discussed in the previous section, as well as for a scenario that corresponds to the evolution of primary production anticipated by Northey *et al.* [NOR 14] (Figures 6.8 and 6.14). In all cases, the annual flow of recycled copper up to 2050 is less than the recoverable copper flow in end-of-life products. The trend then reverses, and the annual production of recycled copper becomes higher than the available flow. This implies that before 2050, the recycling infrastructure (Capital II) is not sufficient to absorb the annual flow of end-of-life products. The available copper that has not yet been recycled is stored in the EOL stock, which peaks at around 150 Mt in 2060. This is the main difference between the model in Figure 6.16 and that in Figure 6.3, where we assumed that the stock of metal to be recycled was kept at zero (recycling of all available copper, without economic constraints). After 2050, Capital II is large enough to absorb not only the annual copper flow in end-of-life products, but also part of the EOL stock that declines over time. The rate of decline of this stock increases with the price of copper subsequent to the collapse of primary production (scenarios 3 and "Northey"). The annual secondary production is proportional to the size of Capital II, which itself grows faster if the price and profits of selling metal increase. The strong growth in annual flow of recycled copper from 2040 onwards is therefore controled by profits from the recycling activity, which are invested to increase Capital II and the size of the production infrastructure.

The model indicates that the sum of primary contributions and recycling allow for a total copper production of over 30 Mt/year between 2020 and 2080 (Northey scenario) or until beyond the end of the century (scenarios 3 and 5). This suggests that future copper supply, if still relevant, cannot be predicted by a simple model that only considers current or estimated geological reserves, as is often the case in the literature (see section 6.2). The future contribution of recycling will be increasingly significant as the amount of metal available in end-of-life products produced 30 years earlier will be greater. However, there are two important limitations on the model proposed in Figure 6.16: first is the assumption that the price of copper is controled by primary production. In reality, there will be competition between the primary sector and the recycling sector as soon as the contribution of recycling is as significant as that of the primary. Lower recycling costs than primary production will contribute to drive the price of copper down, while higher costs of primary production after a possible production peak will have the opposite effect. This situation will naturally be unfavorable to primary production, and likely it will not be able to develop as expected in scenarios 3 and "Northey". Stabilization of primary production, as in scenario 5 (constant price), is possible. It will maintain an increase in the total amount of copper produced at 50 Mt by 2070, which should meet demand. Obviously, these changes are only valid if the type of technologies and exploited deposits remain constant over time. If new primary resources can be exploited at a reasonable cost, the recycling sector could find itself penalized. As already mentioned, another limitation of the model presented in Figure 6.16 is the absence of a true economic model that allows for coupling between demand, production and prices, and takes into account the weight of the debt. In any case, the model results suggest that we should anticipate the transition from a primary production economy to a recycling-oriented economy. Rich countries tend to squander their metal reserves in end-of-life products by exporting them to developing countries. This is contrary to the evolutionary dynamics illustrated in Figure 6.18, which show that the development of a significant recycling activity after 2050 not only requires access to a large flow of end-of-life products, but also to an equally large accessible copper stock (EOL). Indeed, the "EOL" stock controls the growth rate of the recycling infrastructure (Capital II) when economic conditions are met. In the absence of such available stock, the recycling infrastructure cannot be developed locally in an optimal manner.

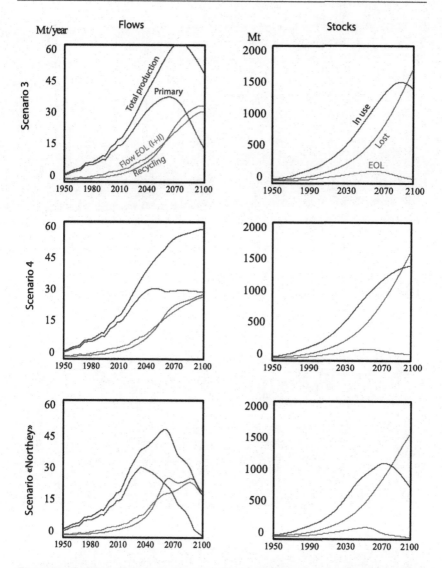

Figure 6.18. *Flows (left) and stocks (right) calculated using the model in Figure 6.16 for scenarios 3 and 4, as discussed in section 6.2.5.2, or to reproduce the evolution of primary production estimated by Northey et al. [NOR 14] at increasing copper prices after the peak of production (2014). For a color version of this figure, see www.iste.co.uk/vidal/energy.zip*

Conclusion

The energy transition will be a source of over-consumption of fossil fuel and energy resources. This is acceptable if the objective is to reduce our CO_2 emissions in the *long term*. However, this over-consumption comes at a time when humankind is already consuming energy and raw materials at a level that has never been seen before, and growth is expected to continue into the future. In this context, an analysis of the needs and impacts related to the rapid evolution of our energy system is vital to avoid mass deployment of technologies that could lead us into a deadlock, even if they seem viable from a technical-economic point of view.

Contribution and limits of models

History shows that the needs and uses of energy and raw materials can change very rapidly, with consequences that are difficult to predict accurately. In the past, the replacement of biomass with fossil fuels helped to preserve and rebuild Europe's overexploited forests. The use of fossil fuel energy also preserved cetaceans, whose fat and spermaceti were used in oil lamps. However, the use of fossil fuels has also led to technological advances that, a few decades later, have made possible the mass felling of primary forests and hunting of whales for other uses. Access to concentrated and easily usable fossil fuel stocks, coupled with rapid technological changes, has also allowed new types of mineral deposits to be exploited, and the reserves of "structural" raw materials have been increasing over time despite increasing consumption. These evolutions were not anticipated by the studies published in the 1970s, many of which predicted resource shortages as we moved into the 21st Century. Of course, this does not mean that we live in a time of abundance, and the sustainability of natural resources exploitation remains more than ever an important question. The only way to obtain answers is through dynamic modeling that combines reserves and primary production, end-of-life quantities of products and recycling, energy, the economy, environmental impacts, as well as changes in technology and demand. Dynamic models can be used to evaluate the raw materials and energy needs, as well as the future of raw materials production and associated environmental impacts. However, the modeling results depend on the input data and assumptions regarding the evolution of all parameters mentioned above. For this reason, evaluating one single accurate and robust future trend based on the sole analysis of historical data of demand, production and primary reserves alone, is clearly impossible. It is only possible to estimate a range of possible futures for assumed evolutions of recycling rates, energy efficiency, economic, social and geopolitical conditions and environmental impacts. Nevertheless, these results are sufficient to identify the less demanding technologies with the lowest impacts, and as illustrated in the previous pages for the energy transition, to establish a quantitative basis to compare the different scenarios. The simplicity of the dynamic models presented in Chapter 6 should not obscure the complexity of real interactions between geological, environmental and economic dimensions. It is certainly possible to dig much deeper into the coupling and links between consumption and environmental impacts. Assessing these impacts usually occurs through life-cycle assesments that consider the detail of the entire value chain in a much more realistic way than the dynamic models used in this book. However, life-cycle assesments have the disadvantage of being static and are based on the current technologies of manufacturing, primary production and recycling, energy costs and impact calculations. In the case of energy scenarios, this approach implies that the production of wind turbines and photovoltaic panels at the end of the scenario have the same environmental impacts as those produced at the beginning. This is not the case in reality, because the energy mix, the energy efficiency, the processes of production and the ratio of primary production to recycling will change over time. An improvement would be to link life-cycle assesments with dynamic models by making assumptions about these changes.

What is the most critical?

Over the past decade, studies and analyses on needs have largely focused on technological metals that show a high current criticality, which is proportional to the product of supply risk multiplied by the vulnerability to supply restriction. However, the use of technological metals is largely dependent on future innovation. Technological metals are often used because they improve the performance of alloys, they reduce the size or weight of components and thus contribute to improving energy efficiency or miniaturization, but they are not absolutely necessary to ensure a function. In addition, the medium-term supply capacities of technological metals are uncertain because knowledge on geological reserves of metals that have been in use for only three decades is poor. For these reasons, a currently high criticality does not have much significance for the next 30 years. On the other hand, structural raw materials do not seem to present any supply risk in the short to medium term (<20 years) because their reserves are huge (in the case of iron, aluminum, carbonates, sand and gravel, etc.). However, their production is equally huge and they are not substitutable, so they are not very dependent on technological innovations. We cannot do without these raw materials, which carry most of the energy used to build future energy infrastructure, as well as the basic infrastructure in our societies, like large urban centers that are springing up in many developing countries. Structural raw materials should therefore not be neglected in the analysis of energy transition scenarios.

Beyond the energy transition

Apart from energy and material needs, the transition to low-carbon and renewable energies raises other questions. Unlike the transition from biomass to fossil fuels, which enabled rapid industrialization in the 20th Century, the transition from fossil fuels (concentrated energy stocks) to renewable energies (diluted energy flows) does not allow for a gain in power or an improvement in ease of use. The compatibility of this transition with economic growth and improvement of the average global standard of living is thus questionable, which was not the case for the transition from biomass to fossil fuel energy. In the short term, it would seem prudent to associate the energy transition with an overall sobriety in energy consumption, as proposed in many scenarios, including the Ecofys-WWF scenario used in the previous pages. Energy efficiency can certainly be improved in all industrial sectors, while people in wealthier countries can be encouraged to adopt more virtuous consumption behaviors. But is it possible to reduce the overall energy consumption while keeping the activities of rich countries at their current level and ensuring growth of poorer countries? The answer is unclear. Chaisson [CHA 14] indicates that ordered "islands of complexity" (low entropy) such as galaxies, stars, planets, the living world and complex societies, appear at all scales in "oceans of disorder" (higher entropy). This author shows that the amount of energy passing through the islands of complexity per unit of mass and time increases with complexity, which itself increases with time. It follows that an overall decrease in energy consumption is impossible in a system of constant size and increasing complexity. In other words, the gain resulting from improved energy efficiency is balanced by the increasing demand in energy accompanying the increase of complexity. Global reduction of energy consumption would be then only possible if the relative size (mass) of the "islands of complexity" decreases compared to the "ocean of disorder". At constant level of complexity, the global reduction of energy consumption would necessitate a decrease in the relative size of urban domains compared to rural areas, or rich countries compared to poor ones. It is therefore unlikely that the global demand for energy will decrease in the future. If we admit that any increase of complexity requires more energy, the limiting factor to growth and developpement is the access to energy. In the long term, the transition from fossil fuels to renewables could help push this limit. Solar radiation comes from outside our Earth system, it is inexhaustible on our time scale and constitutes a vastly greater amount of energy than that used presently by human societies. If technological advances can solve the drawbacks of dilution and intermittency, there is no reason to believe that the overall energy consumption will decrease in the future. The massive switch from fossil fuels to inexhaustible and low-cost renewable energies could even favor increasing energy consumption. In that case, the limiting factor for growth would be the access to non-energy resources and the numerous environmental impacts associated with the increasing metabolism of our societies, which cannot be limited to greenhouse gas emissions. In this perspective and contrary to popular belief, it is not certain that the switch to renewable energies will reduce the environmental pressure in the long term.

The diluted and intermittent nature of renewable energies could also favor the connection between nations and a globalization of energy production and distribution between heavily sunny and windy areas and less windy areas, between night and daylight areas. Trade in energy between countries is not new but a new global system organization will need to be redrawn because future transport vectors will be different, the forms of storage are yet to be invented and primary sources will be located in places different from those of fossil fuels. A global connection allowing exchanges of energy and monetary flows is necessary if we want to maintain a business-as-usual organization and activity of human society that resembles the existing one. However, generating electricity from renewable sources also provides local power. These may be isolated rural areas or small urban areas. This is where the added value of renewable energies fully emerges, as it can allow the emergence of moderately complex "rural" islands without altering "urban" areas. Decentralization of energy production and its local use with limited interconnection can therefore be an alternative, or at least it can be a complementary route to widespread interconnection on a global scale.

All these questions are exciting and must be analyzed by combining the know-how of different actors (scientists, engineers, economists, sociologists, industrialists, politicians, citizens, etc.), because they define the foundations of tomorrow's world, which will undoubtedly be very different from today's world.

Bibliography

[ALI 17] ALI S., GIURCO D., NICKLESS E., *et al.*, "Mineral supply for sustainable development requires resource governance", *Nature*, 543, pp. 367–372, 2017.

[ALL 10] ALLWOOD J.M., CULLEN J.M., MILFORD R.L., "2010 Options for achieving a 50% cut in industrial carbon emissions by 2050", *Environ. Sci. Technol.*, 44, pp. 1888–1894, 2010.

[ALL 12] ALLWOOD J.M., CULLEN J.M., *Sustainable Materials: With Both Eyes Open*, UIT Cambridge, 2012.

[ALO 12] ALONSO E., SHERMAN A.M., WALLINGTON T.J., *et al.*, "Evaluating rare earth element availability: a case with revolutionary demand from clean technologies", *Environ. Sci. Technol.*, 46(6), pp. 3406–3414, 2012.

[ASH 09] ASHBY M.F., *Materials and the Environment: Eco-informed Material Choice*, Butterworth, Burlington, 2009.

[AYR 03] AYRES R.U., AYRES L.W., RÅDE I., "The life-cycle of copper its co-products and byproducts", *Eco-Efficiency in Industry and Science 2003*, 13, 2003.

[AZZ 10] AZZOPARDI B., MUTALE J., "Life cycle analysis for future phtovoltaic system using hybridsolar cells", *Renewable and Sustainable Energy Reviews*, 14, pp. 1130–1134, 2010.

[BAR 09] BARDI U., LAVACCHI A., "A simple interpretation of Hubbert's model of resource exploitation", *Energies*, 2, pp. 646–661, 2009.

[BAS 06] BASALDELLA E.I., PALADINO J.C., VALLE G.M., "Exhausted fluid catalytic cracking catalysts as raw materials for zeolite synthesis", *Applied Catalysis B: Environmental*, 3–4, pp. 189–191, 2006.

[BIR 14] BIRAT J.-P., CHIAPPINI M., RYMAN C. *et al.*, "Cooperation and competition among structural materials", *Revue de Métallurgie*, 110, pp. 97–131, 2014.

[BLE 16] BLEISCHWITZ R., NECHIFOR V., Saturation and Growth Over Time: When Demand for Minerals Peaks, Prisme no. 34, Cournot Centre, 2016. Available at: https://www.centre-cournot.org/img/pdf/prisme_fr/Prisme%20N°34%20Novembre%202 02016%20(english).pdf.

[BOR 01] BORGWARDT R., "Platinum, fuel cells and future US road transport", *Transportation Research, PartD: Transport and Environment*, 6 (3), pp. 199–207, 2001.

[BRA 98] BRANDER J., TAYLOR M.T., "The simple economics of Easter Island: a Ricardo-Malthus model of renewable resource use", *Am. Econ. Rev.*, 88, pp. 119–138, 1998.

[BUR 08] BURFORD B.D., NIVA E., "Comparing Energy Efficiency in Grinding Mills", *Proceedings MetPlant 2008 – Metallurgical Plant Design and Operating Strategies*, Australasian Institute of Mining and Metallurgy, Perth, Australia, pp. 45–64, 2008.

[CHA 74] CHAPMAN P.F., "The energy costs of producing copper and aluminium from primary sources", *Metals and Materials*, 8(2), pp. 107–111, 1974.

[CHA 97] CHASE-DUNN C., HALL T., *Rise and Demise: Comparing World-Systems*, Westview Press, 1997.

[CHA 14] CHAISSON E.J., "The Natural Science Underlying Big History", *The Scientific World Journal*, art. 384912, 2014.

[CLA 09] CLASSEN M., ALTHAUS H.-J., BLASER S., *et al.*, Ecoinvent 2.2: Life Cycle Inventories of Metals, Data v2.0, Ecoinvent Centre, ETh Zurich, 2009.

[COC 14] COMISION CHILENA DEL COBRE (COCHILCO), Statistical database on production and energy use, 2014. Available at: http://www.cochilco.cl/estadisticas/intro-bd.asp.

[CUL 11] CULLBRAND K., MAGNUSSON O., The Use of Potentially Critical Materials in Passenger Cars, Report no. 2012:13, 2011.

[DAL 00] DALTON T.R., COATS R.M., "Could institutional reform have saved Easter Island?", *J. Evol. Econ.*, 10, pp. 489–505, 2000.

[DAL 05] DALTON T.R., COAT R.M., ASRABADI B.R., "Renewable resources, property-rights regimes and endogenous growth", *Ecological Economics*, pp. 52, 31–41, 2005.

[DAS 97] DAS A., KANDPAL T.C., "Iron and Steel manufacturing technologies in India: estimation of CO_2 emission", *International Journal of Energy Research*, 21, p. 1187, 1997.

[DAV 09] DAVIS S., INOGUCHI Y., Marketing Research Report: Zeolites, SRI Consulting, 2009.

[DAV 14] DAVIDSON E., Defining the trend: Cement consumption vs GDP. Global Cement, 2014. Available at: www.globalcement .com/magazine/articles/858-defining-the-trend-cement-consump tion-vs-gdp.

[DEB 13] DE BAKKER J., "Energy use of fine grinding in mineral processing", *Metallurgical and Materials Transactions E*, 1E, pp. 8–19, 2013.

[DEC 05] DECKER C., REUVENY R., "Endogenous technological progress and the malthusian trap: could Simon and Boserup have saved Easter Island?", *Human Ecology*, 33(1), pp. 119–140, 2005.

[ECO 12] ECOFYS-WWF, The Energy Report 100% Renewable Energy by 2050, World Wide Fund for Nature (WWF), Gland, Switzerland, 2012.

[ECO 14] ECOFYS-WWF, Critical materials for the transition to a 100% sustainable energy future, World Wide Fund for Nature (WWF), Gland, Switzerland, 2014.

[FAL 09] FALCONER I., Metals required for the UK's low carbon energy system: The case of copper usage in wind farms, Master's dissertation, University of Exeter, 2009.

[FIZ 15] FIZAINE F., COURT V., "Renewable electricity producing technologies and metal depletion: A sensitivity analysis using the EROI", *Ecological Economics*, 110, pp. 106–118, 20155.

[FRI 11] FRIMMEL H.E., MÜLLER J., "Estimates of Mineral Resource Availability – How Reliable Are They?", *Akad. Geowiss. Geotechn., Veröffentl.*, 28, pp. 39–62, 2011.

[FUJ 05] FUJII H., NAGAIWA T., KUSUNO H., *et al.*, "How to quantify the environmental profile of stainless steel", *SETAC North America 26th Annual Meeting*, Pensacola, 2005.

[GAR 12] GARCIA-OLIVARÈS A., BALLABRERA-POY J., GARCIA-LADONA E., "A global renewable mix with proven technologies and common materials", *Energy Policy*, 41, pp. 561–574, 2012.

[GER 08] GERST M. D., "Revisiting the cumulative grade-tonnage relationship for major copper ore types", *Econ. Geol. 2008*, 103 (3), pp. 615–628, 2008.

[GLÖ 13] GLÖSER S., SOULIER M., TERCERO ESPINOZA L.A., "Dynamic Analysis of Global Copper Flows. Global Stocks, Postconsumer Material Flows, Recycling Indicators, and Uncertainty Evaluation", *Environ. Sci. Technol*, 47, pp. 6564–6572, 2013.

[GOE 08] GOESSLING-REISEMANN S., "Entropy analysis of metal production and recycling", *Management of Environmental Quality. An International Journal*, 19(4), pp. 487–492, 2008.

[GOO 67] GOODWIN R., "A Growth Cycle", in FEINSTEIN C.H. (ed.), *Socialism, Capitalism and Economic Growth*, Cambridge University Press, 1967.

[GOO 10] GOONAN T.G., Copper recycling in the United States in 2004, USGS Circular 1196-X, 2010.

[GRÄ 04] GRÄDEL T.E., VAN BEERS D., BERTRAM M., *et al.*, "The mulilevel cycle of anthropogenic copper", *Environ. Sci. Technol.*, 38, pp. 1253–1261, 2004.

[GRÄ 10] GRÄDEL T.E., CAO J., "Metal spectra as indicators of development", *PNAS*, 107(49), pp. 20905–20910, 2010.

[GRÄ 11a] GRÄDEL T.E., ALLWOOD J., BIRAT J.-P., *et al.*, What Do We Know About Metal Recycling Rates?, USGS Staff – Published Research, Paper 596. http://digitalcommons.unl.edu/usgsstaffp ub/596, 2011.

[GRÄ 11b] GRÄDEL T.E., "On the Future Availability of the Energy Metals", *Annual Review of Materials Science*, vol. 41, pp. 323–335, 2011.

[GRI 08] GRIMES S., DONALDSON J., GOMEZ G.C., Report on the Environmental Benefits of Recycling, Bureau of International Recycling, Imperial College, London, 2008. Available at: https://cari-acir.org/wp-content/uploads/2014/08/BIR_CO2_report.pdf.

[GUT 08] GUTOWSKI T.G., "Thermodynamics and Recycling, A Review", *IEEE International Symposium on Electronics and the Environment*, San Francisco, USA, May 19–20, 2008.

[GUT 11] GUTFLEISCH O., WILLARD M.A., BRÜCK E., *et al.*, "Magnetic materials and devices for the 21st century: stronger, lighter, and more energy efficient", *Adv Mater.*, 23(7), pp. 821–42, 2011.

[GUT 12] GUTOWSKI T.G., SAHNI S., ALLWOOD J.M., "The energy required to produce materials: constraints on energy-intensity improvements, parameters of demand", *Philosophical Transactions of the Royal Society A.*, 2012. Available at: http://dx.doi.org/10.1098/rsta.2012.0003.

[HAR 10] HARRISON G.P., MACLEAN E.J., KARAMANLIS S., *et al.*, "Life cycle assessment of the transmission network in Great Britain", *Energy Policy*, 38, pp. 3622–3631, 2010.

[HER 15] HERTWICH E.G., GIBON T., BOUMAN E.A., "Integrated life-cycle assessment of electricity-supply scenarios confirms global environmental benefit of low-carbon technologies", *PNAS*, 112(20), pp. 6277–6282, 2015.

[HUB 56] HUBBERT M.K., "Nuclear Energy and Fossil Fuels", *Drilling and Production Practice*, 23, pp. 7–25, 1956.

[HUB 62] HUBBERT M.K., Energy Resources: A Report to the Committee on Natural Resources. Publication 1000-D, National Academy of Sciences – National Research Council, Washington, 1962.

[HUB 82] HUBBERT M.K., Techniques of Prediction as Applied to Production of Oil and Gas, U.S. Department of Commerce, NBS Special Publication, 631, pp. 16–141, 1982.

[IEA 10a] INTERNATIONAL ENERGY AGENCY (IEA), World Energy Outlook 2010, Paris, France, 2010.

[IEA 10b] INTERNATIONAL ENERGY AGENCY (IEA), Energy Technology Perspectives 2010: Scenarios and Strategies to 2050. Paris, France, 2010. Available at: http://www.iea.org/publications/freepublications/publication/name,26100,en.html.

[IEA 13] INTERNATIONAL ENERGY AGENCY (IEA), International Energy Outlook 2013, US Energy Information Administration, Washington, USA. Available at: https://www.eia.gov/outlooks/ieo/pdf/0484(2013).pdf.

[IEA 16] INTERNATIONAL ENERGY AGENCY (IEA), International Energy Outlook 2016. US Energy Information Administration, Washington, USA. Available at: https://www.eia.gov/outlooks/ieo/pdf/0484(2016).pdf.

[JOH 07] JOHNSON J., HARPER E.M., LIFSET R. et al., "Dining at the periodic table: metal concentrations as they relate to recycling", Environ. Sci. Technol., 41, pp. 1759–1765, 2007.

[JOH 14] JOHNSON K.M., HAMMARSTROM J.M., ZIENTEK M.L., et al. Estimate of undiscovered copper resources of the world, 2013, U.S. Geological Survey Fact Sheet 2014–3004, 2014. Available at: http://dx.doi.org/10.3133/fs20143004.

[KER 14] KERR R.A., "The coming copper peak", Science, 343, pp. 722–724, 2014.

[KLE 11] KLEIJN R., VAN DER VOET E., KRAMER G.J., et al., "Metal requirements of low-carbon power generation", Energy, 36(9), 5640–5648, 2011.

[KLE 12] KLEIJN R., Materials and energy: a story of linkages, material requirements of new energy technologies, resource scarcity and interconnected material flows, PhD thesis, Leiden University, Netherlands, 2012.

[KLO 08] KLOBASA M., SENSFUSS F., ERGE T., et al., "Analysis of the contribution of load management to the cost-efficient balancing of wind energy and the mitigation of grid congestions", DEWEK 2008, 9th German Wind Energy Conference. Proceedings, 2008.

[KOL 10] KOLTUN P., THARUMARAJAH A., "LCA study of rare earth metals for magnesium alloy applications", *Materials Science Forum*, 654–656(1), pp. 803–806, 2010.

[LAH 10] LAHERRÈRE J., "Copper peak", *The Oildrum Europe*, 630, pp. 1–27, 2010. Available at: http://europe.theoildrum.com/node/6307.

[LOT 25] LOTKA A.J., *Elements of Physical Biology*, Williams and Wilkins, 1925.

[LOT 11] LOTTERMOSER B.G., "Recycling, Reuse and Rehabilitation of Mine Wastes", *Elements*, 7, pp. 405–410, 2011.

[LOT 12] LOTTERMOSER B.G., *Mine Wastes: Characterization, Treatment and Environmental Impacts*, Springer, New York, 2012.

[MAL 13] MALVOISIN B., BRUNET F., CARLUT J., "High-purity hydrogen gas from the reaction between BOF steel slag and water in the 473–673 K range", *International Journal of Hydrogen Energy*, 38(18), pp. 7382–7393, 2013.

[MAR 06] MARTYUSHEV L. M., SELEZNEV V.D., "Maximum entropy production principle in physics, chemistry and biology", *Physics Reports*, 426(1), pp. 1–45, 2006.

[MAS 06] MASON J.M., FTHENAKIS V.M., HANSEN T., *et al.*, "Energy Pay-Back and Life Cycle CO2 Emissions of the BOS in an Optimized 3.5 MW PV Installation", *Prog. In Photovoltaics Res. and Applications*, 14, pp. 179–190, 2006.

[MAX 00] MAXWELL J.W., REUVENY R., "Resource scarcity and conflict in developing countries", *Journal of Peace Research*, 37(3), pp. 301-322, 2000.

[MEA 72] MEADOWS D.H., MEADOWS D.L., RANDERS J. *et al.*, *The Limits to Growth*, Universe Books, 1972.

[MEI 16] MEINERT L.D., ROBINSON G.R., NASSAR N.T., "Mineral Resources: Reserves, Peak Production and the Future", *Resources* 5, p. 14, 2016.

[MOR 11] MORGAN J.P., Rare Earths. We touch them every day, The Global Business Dialogue, 2011. Available at: http://www.gbdinc.org/PDFs/RARE%20EARTH%20PAPER%20FROM%20T.%20STEWART%20JUNE%209%202011.pdf.

[MOS 13] Moss R.L., Tzimas E., Willis P., *et al.*, Critical metals in the path towards the decarbonisation of the EU energy sector, Assessing rare metals as supply-chain bottlenecks in low-carbon energy technologies, JERC Pub. No JRC65592, EUR 24884 EN, Joint Research Centre, 2013. Available at: http://publications. jrc.ec.europa.eu/repository/handle/111111111/22726.

[MOT 14] Motesharrei S., Rivas J., Kalnay E., "Human and nature dynamics (HANDY): Modeling inequality and use of resources in the collapse or sustainability of societies", *Ecological Economics*, 10, pp. 90–102, 2014.

[MUD 07] Mudd G., "An analysis of historic production trends in Australian base metal mining", *Ore Geology Reviews*, 32, pp. 227–261, 2007.

[MUD 08] Mudd G., Diesendorf M., "Sustainability of uranium mining and milling: toward quantifying resources and eco-efficiency", *Environ. Sci. Technol.*, 42(7), pp. 2624–30, 2008.

[MUD 10] Mudd G., "The environmental sustainability of mining in Australia: keymega-trends and looming constraints", *Resources Policy*, vol. 35, no. 2, pp. 98–115, 2010.

[MUD 13a] Mudd G., "The 'Limits to Growth' and 'Finite' Mineral Resources - Re-visiting the assumptions and drinking from that half-capacity glass", *Int. J. Sustainable Development*, 16 (3/4), p. 204, 2013.

[MUD 13b] Mudd G., Weng Z., Jowitt S. M., "A Detailed Assessment of Global Cu Resource Trends and Endowments", *Economic Geology*, 108, pp. 1163–1183, 2013.

[MÜL 11] Müller D.B., Wang T., Duval B., "Patterns of iron use in societal evolution", *Environ. Sci. Technol.*, 45, pp. 182–188, 2011.

[NOR 00] Norgate T.E., Rankin W.J., "Life cycle assessment of copper and nickel production", *MINPREX 2000 International Congress on Mineral Processing and Extractive Metallurgy*, Melbourne, pp. 133–138, 11–13 September 2000.

[NOR 02a] Norgate T.E., Rankin W.J., "The role of metals in sustainable development", *Green Processing*, Australasian Institute of Mining and Metallurgy, Melbourne, pp. 49–55, 2002.

[NOR 02b] NORGATE T.E., RANKIN W.J., "An environmental assessment of lead and zinc production", *Green Processing*, Australasian Institute of Mining and Metallurgy, Melbourne, pp. 177–184, 2002.

[NOR 07] NORGATE T. E., JAHANSHAHI S., RANKIN W. J., "Assessing the environmental impact of metal production processes", *Journal of Cleaner Production* 15(8–9), pp. 838-848, 2007.

[NOR 10a] NORGATE T.E., JAHANSHAHI S., "Low grade ores. Smelt, leach or concentrate?", *Miner. Eng.*, 23, pp. 65–73, 2010.

[NOR 10b] NORGATE T.E., "Deteriorating ore resources: energy and water impacts", in GRAEDEL T., VAN DER VOET E. (eds), *Linkages of Sustainability*, MIT Press, Cambridge, 2010.

[NOR 10c] NORGATE T.E., "Energy and greenhouse gas impacts of mining and mineral processing operations", *Journal of Cleaner Production*, 18(2010), pp. 266–274, 2010.

[NOR 14] NORTHEY S., MOHR S., MUDD G.M., *et al.*, "Modelling future copper ore grade decline based on a detailed assessment of copper resources and mining", *Resources, Conservation and Recycling*, 83, pp. 190–211, 2014.

[NUS 14] NUSS P., ECKELMAN M., "Life Cycle Assessment of Metals: A Scientific Synthesis", *PLOS ONE*, 9(7), 2014.

[ÖHR 12] ÖHRLUND I., STOA, Science and Technology Options Assessment, Future Metal Demand from Photovoltaic Cells and Wind Turbines, Investigating the Potential Risk of Disabling a Shift to Renewable Energy, Science and Technology Options Assessment (STOA), Brussels, 2012. Available at: http://go.nature.com/VUOs7V.

[OKU 15] OKULLO S.J., REYNÈS F., HOFKES M.W., "Modeling peak oil and the geological constraints on oil production", *Resource and Energy Economics*, 40, pp. 36–56, 2015.

[PAC 02] PACCA S., HORVATH A., "Greenhouse Gas Emissions from Building and Operating Electric Power Plants in the Upper Colorado River Basin", *Environ. Sci. Technol.*, 36, pp. 3194–3200, 2002.

[PER 09] PERPINAN O., LORENZO E., CASTRO M.A., et al., "Energy Payback time of grid connected PV systems: comparison between tracking and fixed systems", *Progress in Photovoltaics: Research and Applications*, 17, pp. 137–147, 2009.

[PEZ 03] PEZZEY J.C.V., ANDERIES J.M., "The Effect of Subsistence on collapse and institutional adaptation in population-resource societies", *Journal of Development Economics*, 72(1), pp. 299–320, 2003.

[PHI 76] PHILLIPS W.G.B., EDWARDS D.P., "Metal prices as a function of ore grade", *Resources Policy*, 2(3), pp. 167–178, 1976.

[RAB 12] RABINOVITCH M., "Perspectives de la géologie minière européenne", *Geologues*, 153, pp. 4–22, 2012.

[RAH 11] RAHIMI N., KARIMZADEH R., "Catalytic cracking of hydrocarbons over modified ZSM-5 zeolites to produce light olefins: A review", *Applied Catalysis A: General*, 398, 1-2, pp. 1–17, 2011.

[RAJ 05] RAJARAM R., MELCHERS R.E., "Waste Management", in RAJARAM V., DUTTA S. (eds), *Sustainable Mining Practices - A Global Perspective*, A. A. Balkema Publishers, Leiden, Netherlands, 2005.

[RAN 11] RANKIN W.J., *Minerals, Metals and Sustainability: Meeting Future Material Needs*, CSIRO Publishing, Collingwood, Australia, 2011.

[RAN 12] RANKIN W.J., "Energy Use in Metal Production", *High Temperature Processing Symposium 2012*, Swinburne University of Technology, 2012. Available at: https://publications.csiro.au/rpr/download?pid=csiro:EP12183&dsid=DS3

[RAU 07] RAUGEI M. BARGIGLI S., ULGIATI S., "Life cycle assessment and energy pay-back time of advanced photovoltaic modules: CdTe and CIS compared to poly-Si", *Energy*, 32, pp. 1310–1318, 2007.

[RAU 09] RAUCH J. N., "Global mapping of Al, Cu, Fe, and Zn in-use stocks and in-ground resources", *Proceedings of the National Academy of Sciences of the United State of America*, 106(45), pp. 18920–18925, 2009.

[REI 13] REICHL C., SCHATZ M., ZSACK G., World Mining Data (2013), Bundesministerium für Wirtschaft und Arbeit, Vienna, Austria, 2013. Available at: http://www.wmc.org.pl/?q=node/49.

[REU 00] REUVENY R., DECKER C.S., "Easter Island: historical anecdote or warning for the future?", Ecol. Econ., 35, pp. 271–287, 2000.

[REY 10] REYNÈS F., OKULLO S., HOFKES M., How Does Economic Theory Explain the Hubbert Peak Oil Model?, USAEE Working Paper No. 10–052, 2010. Available at: http://ssrn.com/abstract=1711610.

[ROS 76] ROSENKRANZ R.D., Energy consumption in domestic primary copper production, Information circular 8698, U.S. Bureau of Mines, Washington D.C., 1976. Available at: https://ia600801.us.archive.org/26/items/energyconsumptio00ros e/energyconsumptio00rose.pdf.

[RUL 09] RULE B.M., WORTH Z.J., BOYLE C.A., "Comparison of Life Cycle Carbon Dioxide Emissions and Embodied Energy in Four Renewable Electricity Generation Technologies in New Zealand", Environ. Sci. Technol., 43, pp. 6406–6413, 2009.

[SAF 07] SAFIROVA E., HOUDE S., HARRINGTON W., Spatial Development and Energy Consumption, Resources for the Future, Washington D.C., December 2007. Available at: http://www.rff.org/files/sharepoint/WorkImages/Download/RFF-DP-07-51.pdf.

[SAT 11] SATHAYE J., LUCON O., RAHMAN A., et al., "Renewable Energy in the Context of Sustainable Development", in EDENHOFER O., PICHS-MADRUGA R., SOKONA Y., et al. (eds), IPCC Special Report on Renewable Energy Sources and Climate Change Mitigation, Cambridge University Press, 2011.

[SCH 10] SCHODDE R., "The key drivers behind resource growth: an analysis of the copper industry over the last 100 years", 2010 MEMS Conference Mineral and Metal Markets over the Long Term. SME Annual Meeting, Phoenix, March 3 2010. Available at: http://www.minexconsulting.com/publications/Growth%20Fa ctors%20for%20Copper%20SME-MEMS%20March%202010.pdf.

[SIL 12] SILVA A.C., BERNADES A.T., LUZ J.A.M., "Simulation of the mineral breakage using a fractal approach", *R. Esc. Minas, Ouro Preto*, 65(2), pp. 285–288, 2012.

[STE 14] STEFFEN W., BROADGATE W., DEUTSCH L., *et al.*, "The Trajectory of the Anthropocene: The Great Acceleration", *The Anthropocene Review*, Vol. 2(1) pp. 81–98, 2014.

[SVE 14a] SVERDRUP H.U., RAGNASDOTTIR K.A., "Natural resources in a planetary prespective", *Geochemical Perspectives*, 3(2), 2014.

[SVE 14b] SVERDRUP H.U., RAGNASDOTTIR K.A., KOCA D., "On modelling the global copper mining rates, market supply, copperprice and the end of copper reserves", *Resources, Conservation and Recycling*, 87(2014), pp. 158–174, 2014.

[TAH 97] TAHARA K., KOJIMA T., INABA A., "Evaluation of CO2 Payback Time of Power Plants by LCA", *Energy Convers. Manage*, 38(Suppl.), pp. 615-620, 1997.

[TAI 88] TAINTER J.A., *The Collapse of Complex Societies*, Cambridge University Press, 1988.

[TAL 11] TALENS PEIRO L., VILLALBA MENDEZ G., AYRES R., Rare and Critical Metals by-products and the implication for future supply, Faculty and Research Working Paper, 2011. Available at: http://www.insead.edu/facultyresearch/research/doc.cfm?did=48 916.

[TIL 07] TILTON J.E., LAGOS G., "Assessing the long-run availability of copper", *Resources Policy*, 32, pp. 19–23, 2007.

[TUR 09] TURCHIN P., NEFEDOV S.A., *Secular Cycles*, Princeton University Press, 2009.

[UNE 10a] UNEP, Assessing the Environmental Impacts of Consumption and Production: Priority Products and Materials, A Report of the Working Group on the Environmental Impacts of Products and Materials to the International Panel for Sustainable Resource Management, 2010. Available at: http://www.unep.org/resourcepanel/Portals/24102/PDFs/Priority ProductsAndMaterials_Report.pdf.

[UNE 10b] UNEP, Metal stocks in society – A Report of the Working Group on the Global Metal Flows to the International Resource Panel, 2010. Available at: http://www.unep.fr/shared/publications/pdf/DTIx1264xPA-Metal%20stocks%20in%20society.pdf.

[UNE 11] UNEP, Recycling rates of metals - A Status report. A Report of the Working Group on the Global Metal Flows to the UNEP International Resource Panel, 2011. Available at: http://www.unep.org/resourcepanel/Portals/24102/PDFs/Metals_Recycling_Rates_110412-1.pdf.

[UNE 13a] UNEP, Environmental Risks and Challenges of Anthropogenic Metals Flows and Cycles, A Report of the Working Group on the Global Metal Flows to the International Resource Panel, 2013. Available at: http://www.unep.org/resourcepanel/Portals/24102/PDFs/Environmental_Challenges_Metals-Full%20Report_150dpi_130923.pdf.

[UNE 13b] UNEP, Metal Recycling: Opportunities, Limits, Infrastructure, A Report of the Working Group on the Global Metal Flows to the International Resource Panel, 2013.

[USD 11] U.S. DEPARTMENT OF ENERGY, Critical materials strategy, Washington D.C., 2011. Available at: http://energy.gov/sites/prod/files/DOE_CMS2011_FINAL_Full.pdf.

[VAL 13] VALERO A., VALERO, A., DOMÍNGUEZ A. "Exergy Replacement Cost of Mineral Resources", *Journal of Environmental Accounting and Management*, 1(2), pp. 147–168, 2013.

[VAL 15] VALERO A., VALERO A., "Thermodynamic Rarity and the Loss of Mineral Wealth", *Energies*, 8, pp. 821–836, 2015.

[VID 13] VIDAL O., GOFFÉ B., ARNDT N., "Metals for a low-carbon society", *Nature Geoscience*, 6, pp. 894–896, 2013.

[VIE 12] VIEIRA M.D.M, GOEDKOOP M.J., STORM P. *et al.*, "Ore Grade Decrease as Life Cycle Impact Indicator for Metal Scarcity: The Case of Copper", *Environ. Sci. Technol.*, 46, pp. 12772–12778, 2012.

[VOL 26] VOLTERRA V., "Variazioni e fluttuazioni del numero dindividui in specie animali conviventi", *Memoria Accademia dei Lincei Roma*, 2:31(113), 1926.

[WIE 15] WIEDMANN T.O., SCHANDL H., LENZEN M., *et al.*, "The material footprint of nations", *PNAS*, 112(20), pp. 6271–6276, 2015.

[YEL 10] YELLISHETY M., RANJITH P.G., THARUMARAJAH A., "Iron ore and steel production trends and material flows in the world: Is this really sustainable?", *Resources, Conservation and Recycling*, 54, pp. 1084-1094, 2010.

[YEN 07] YENTEKAKIS I.V., KONSOLAKIS M., RAPAKOUSIOS I.A., *et al.*, "Novelelectropositively promoted monometallic (Pt-only) catalytic converters for automotive pollution control", *Topics in Catalysis*, 42-43, pp. 393–397, 2007.

[YOU 02] YOUNGER P.L., BANWART S.A., HEDIN R.S., *Mine Water: Hydrology, Pollution, Remediation,* Kluwer Academic Publishers, Dordrecht, 2002.

[ZEP 14] ZEPF V., RELLER A., RENNIE C., *et al.*, *Materials Critical to the Energy Industry. An Introduction*, 2nd edition, BP, 2014. Available at: http://www.bp.com/energysustainabilitychallenge.

Index

Printed in the United States
By Bookmasters